SHIFTING SHORES:

marsh expansion and retreat in San Pablo Bay

AUTHORS

Julie Beagle

Micha Salomon

CONTRIBUTING AUTHORS

Robin Grossinger

Sean Baumgarten

DESIGN AND LAYOUT ·

Ruth Askevold

SFEI
AOSOC

PREPARED BY San Francisco Estuary Institute-Aquatic Science Center

IN COOPERATION WITH AND FUNDED BY S.F. Bay Water Quality Fund,
U.S. Environmental Protection Agency

and the
San Francisco Estuary Partnership

SFEI-ASC PUBLICATION #751   June 2015

SUGGESTED CITATION

Beagle JR, Salomon M, Baumgarten SA, Grossinger RM. 2015. Shifting shores: Marsh expansion and retreat in San Pablo Bay. Prepared for the US EPA San Francisco Bay Program and the San Francisco Estuary Partnership. A Report of SFEI-ASC's Resilient Landscapes Program, Publication # 751, San Francisco Estuary Institute, Richmond, CA.

Version: 1.1 (minor edits made to document and printed May 2017)

REPORT AVAILABILITY

Report is available on SFEI's website at www.sfei.org/projects/shorelinechange.

COVER CREDITS

Photo by Micha Salomon, February 2015.

CONTENTS

ACKNOWLEDGMENTS

This project was funded by the United States Environmental Protection Agency (USEPA) San Francisco Bay Program through the Estuary 2100 Phase 2 grant program. Many thanks to the San Francisco Estuary Partnership for managing the grant.

The project has benefited substantially from the sound technical guidance, engagement, and enthusiasm contributed by several advisors: Jeremy Lowe (formerly with ESA-PWA, currently with SFEI-ASC), Peter Baye (coastal ecologist), Glenn Guntenspergen (USGS), and John Callaway (USF). Josh Collins and Robin Grossinger (SFEI-ASC) were instrumental in guiding and shaping this project from start to finish.

EXECUTIVE SUMMARY

As sea level rise accelerates, our shores will be increasingly vulnerable to erosion. Particular concern centers around the potential loss of San Francisco Bay's much-valued tidal marshes, which provide natural flood protection to our shorelines, habitat for native wildlife, and many other ecosystem services. Addressing this concern, this study is the first systematic analysis of the rates of marsh retreat and expansion over time for San Pablo Bay, located in the northern part of San Francisco Bay.

Key findings:

- Over the past two decades, more of the marshes in San Pablo Bay have expanded (35% by length) than retreated (6%).

- Some areas have been expanding for over 150 years.

- Some marsh edges that appear to be retreating are in fact expanding rapidly at rates of up to 8 m/yr.

- Marsh edge change may be a useful indicator of resilience, identifying favorable sites for marsh persistence.

- These data can provide a foundation for understanding drivers of marsh edge expansion and retreat such as wind direction, wave energy, watershed sediment supply, and mudflat shape.

- This understanding of system dynamics will help inform management decisions about marsh restoration and protection.

- This study provides a baseline and method for tracking marsh edge response to current and future conditions, particularly anticipated changes in sea level, wave energy, and sediment supply.

Recommended next steps:

- This pilot study for San Pablo Bay marshes should be extended to other marshes in San Francisco Bay.

- These initial marsh expansion and retreat findings should be further analyzed and interpreted to improve our understanding of system drivers and identify management responses.

- A program for repeated assessment should be developed to identify and track changes in shoreline position, a leading indicator of the likelihood marsh survival.

GLOSSARY

shoreline definition

Bayward edge of the marsh plain (marsh edge)

For the purposes of this project we are measuring the changes of a geomorphic transition between mid-marsh and low marsh, the bayward edge of the marsh plain. This edge separates the bayward, unconsolidated low marsh (dominated by cordgrass [*Spartina* spp.] and mostly unvegetated mudflats) from the landward, consolidated, mostly vegetated mid-marsh (dominated by pickleweed [*Sarcocornia pacifica*]).

Shoreline (coastline)

The intersection of the land with the water surface. The shoreline shown on charts represents the line of contact between the land and a selected water elevation. In areas affected by tidal fluctuations, this line of contact is the mean high water (MHW) line. In confined coastal waters of diminished tidal influence, the mean water level line may be used (NOS CO-OPS 1 2000).

shoreline movement

Expansion

Continuing bayward movement of the shoreline; advancing edge of marsh, a net bayward movement of the shoreline over a specified period. Also known as progradation.

Retreat

Continuing landward movement of the shoreline; a net landward movement of the shoreline over a specified period. Also known as lateral erosion, or recession.

Accretion

The gradual and imperceptible vertical accumulation of land by natural causes. This may be the result from a deposit of alluvium upon the shore, or by a recession of the water from the shore. Most often used when describing changes in elevation of marshes (Shalowitz 1964).

Transgression

The movement of marsh (or other) habitat upslope as sea levels rise. Depends on the availability of land and the availability of slopes low enough for marshes to form. Marshes are squeezed, or compressed, when they are unable to transgress upslope (Goals Project 2015).

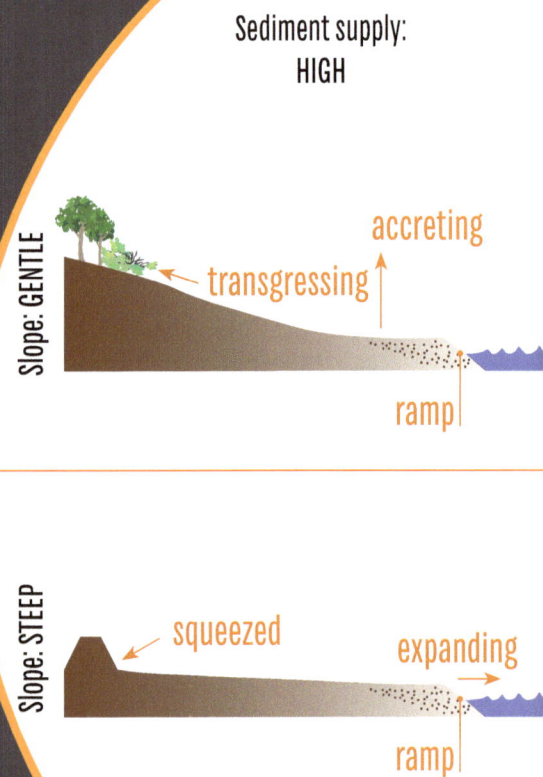

Sediment supply: HIGH

Slope: GENTLE — transgressing — accreting — ramp

Slope: STEEP — squeezed — expanding — ramp

Sediment supply:
LOW

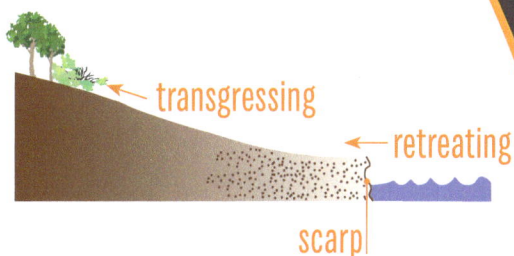

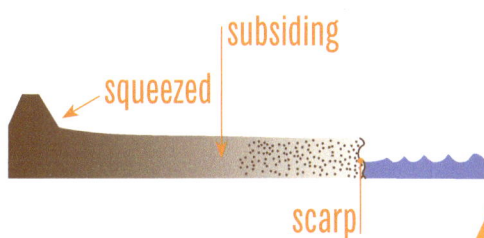

transgressing

retreating

scarp

subsiding

squeezed

scarp

Scarp

Vertical marsh face between the salt marsh and the tidal flat. Edge of the marsh plain (1.2-2.2 m in height) as measured in San Pablo Bay.

Bluff

A cliff or headland with an almost perpendicular face (Hydrographic Dictionary 1990). [Cliff: Land rising abruptly for a considerable distance above the water or surrounding land. (Hydrographic Dictionary 1994)].

Ramp

A marsh face that is sloping, or in an inclined position normal to wave attack.

Berm (wave-built)

Nearly horizontal portion of a beach or backshore having an abrupt fall and formed by wave deposition of material and marking the limit of ordinary high tides (Ellis 1978). A portion of marsh that is higher than the surrounding marsh plain, formed by wave deposition of material.

Baylands

General term describing areas around the margin of a bay, including mudflats, tidal marsh, transition zone (Goals Project 2015).

landscape features

ADAPTED FROM BRINSON ET AL. 1995

Petaluma

Novato

SAN
PABLO
BAY

Vallejo

Martinez

San Rafael

Richmond

Berkeley

Sausalito

SAN

OAKLAND

SAN
FRANCISCO

FRANCISCO

Pleasant

Hayward

BAY

San Mateo

Fremont

N

1:350,000

5 miles

Mountain
View

SAN
JOSE

1. INTRODUCTION

San Francisco Bay is home to the majority of California's tidal marshlands. These marshes provide essential habitat for endangered species, shorebirds, and waterfowl on the Pacific Flyway, and a range of additional fish and wildlife species (Goals Project 1999). In addition to their ecological significance, the marshes provide important ecological services to the region, including flood protection capacity, storm surge buffers (Cooper et al. 2001, BCDC 2013), carbon storage, and chemical/physical filtration of urban and agricultural storm waters (Odum 1990, Goals Project 2015). However, accelerated sea level rise associated with climate change threatens the survival of San Francisco Bay's valued marshes in coming decades. Coastal erosion, in particular, has become an increasingly significant issue for the marshes that ring the San Francisco Bay.

Tidal marshes are dynamic and constantly evolving. Some marsh surfaces are increasing in elevation due to accretion of organic and inorganic sediments; some are decreasing in elevation through erosion, subsidence, organic decomposition, or autocompaction. Some areas of marsh are retreating (eroding horizontally), while others are laterally expanding bayward (prograding) or moving upslope (transgressing). The direction and rate of these changes are driven by varied physical and ecological factors, such as elevation with respect to the tide, orientation, wave energy, vegetation type, shoreline structure, sediment supply, and land availability (Schwimmer and Pizzuto 2000).

As sea level rises, the tidal marshlands will continue to evolve in three major directions: vertically accreting or downshifting (depending on sediment supply and organic accumulation), migrating upslope and inland (depending on accommodation space), and laterally expanding or retreating at the Bay edge (Brinson et al. 1995). While vertical marsh elevation changes have been studied in San Francisco Bay (Patrick and DeLaune 1990, Goals Project 1999, Strahlberg et al. 2011, Swanson et al. 2014), less attention has been paid to the dynamics of the bayward edge of the marsh plain (the "marsh edge"). Lateral changes in the position of

Livermore

San Pablo Bay. (imagery courtesy ESRI)

3

the marsh edge are extremely important because marsh retreat is thought to be the chief mechanism by which coastal wetlands worldwide are being lost (Francalanci et al. 2011, Marani et al. 2011, Fagherazzi 2013). This component of marsh dynamics has not yet been systematically analyzed in San Francisco Bay.

Accurate measurements of the marsh edge will be a critical dataset for managing marshes in the coming decades. These data provide an indication of shoreline resilience, serve as an input for sea level rise response models, and can help in prioritizing restoration and adaptation strategies. In light of accelerated sea level rise and changing sediment availability, without this basic understanding of shoreline dynamics the region may expend valuable resources in unsustainable places.

In this study, we used a systematic, empirical, and repeatable approach to map the bayward marsh edge around San Pablo Bay (in the northern part of San Francisco Bay) for three time periods: ca. 1855, 1993, and 2010. We then quantified changes in marsh edge position to identify zones of expansion and retreat. This report aims to increase our understanding of the rate, distribution, and mechanisms of shoreline change over the long- and short-term, and to provide recommendations for tracking this change in the future. The results of this pilot study provide a new level of understanding about the dynamics of our marsh shorelines and the ways they are likely to respond to local actions.

Examples of the diverse shoreline of San Pablo Bay. The San Pablo Bay shoreline is predominately fronted by extant tidal marsh, and is punctuated with rocky man-made protrusions, some headlands, and beaches. The marsh edges vary in morphology, as exemplified by the single-ridge berm along the Marin County shoreline near Hamilton Wetlands **(top left)**, the scarps in Point Pinole in Contra Costa County **(bottom left)**, and the low gradient ramped morphology in Sonoma County along Highway 37 **(right)**. ([top left] photo by Julie Beagle, June 2015; [bottom left] photo by Shira Bezalel, February 2015; [right] photo by Micha Salomon, April 2012)

Tidal marshes make up almost 30% of the San Francisco Bay shoreline.
Of that third, roughly 80% is found in San Pablo Bay.

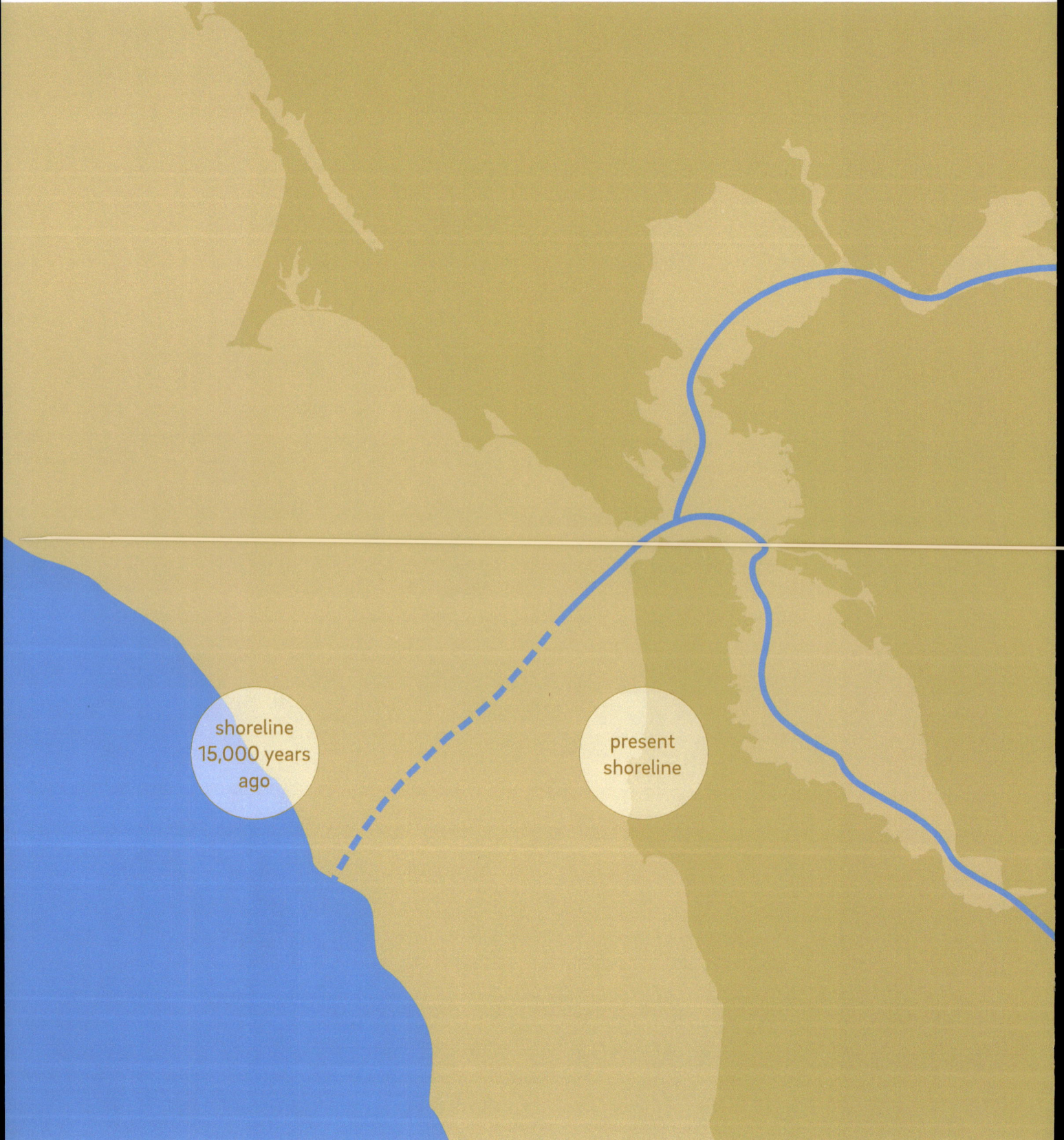

shoreline
15,000 years
ago

present
shoreline

2. BACKGROUND

The shoreline has always been shifting. Around 15,000 years ago the shoreline of the California coast was located west of the Farallon Islands and was 140 m below its current level (left; Cohen and Laws 1992, Malamud-Roam et al. 2007). Rapid sea level rise (estimated at 2 cm/yr [0.8 in/yr or 80 in/century]) following the end of the last glacial epoch led to the inundation of what is now San Francisco Bay. By approximately 6,000 years ago the rate of sea level rise had slowed to 0.25 cm/yr (1 in/yr or 10 in/century), allowing broad tidal marshes to form and maintain themselves (Atwater 1979, Goals Project 1999, Mudd 2011). During the 1850s, many marshes expanded extensively due to increased sediment supply from hydraulic mining in the Sierras (Gilbert 1917, Goals Project 1999, Kirwan et al. 2011), which ushered in a period of relative stasis in marsh extent around the Bay when sea level rise was about 0.2 cm/ yr (0.07 in/yr or 7 in/century) (Malamud-Roam and Ingram 2004). Currently, however, a combination of decreased sediment supply, increased rates of relative sea level rise (between 91 and 140 cm/ century or 36 and 55 in/century), and human modifications along the shoreline threaten the resilience of the San Francisco Bay marshes.

Shoreline change over 15,000 years. Since the last ice age, the seas have been moving steadily up and inland. The rate of advance inland along marsh edges is mediated by local factors such as sediment supply, shoreline modification, and hydrodynamics. (adapted from Cohen and Laws 1992)

Setting

The study area for this project encompasses the bayward edge of the marshes of San Pablo Bay, from Point San Pedro in eastern Marin County to Point San Pablo in western Contra Costa County (below). San Pablo Bay is a pentagonal-shaped embayment in the northern part of the San Francisco Estuary. Drainage from the Central Valley, several sizeable rivers and creeks (including Las Gallinas, Novato, Petaluma, Sonoma, Napa, and Wildcat), and many smaller streams enter San Pablo Bay, bringing sediment and freshwater to the marshes, mudflats, and deep water channels of the Bay. The local watersheds of San Pablo Bay are generally steep with naturally large sediment yields due to friable geology, and tectonic uplift, augmented by accelerated channel incision due to increases in runoff (McKee et al. 2013).

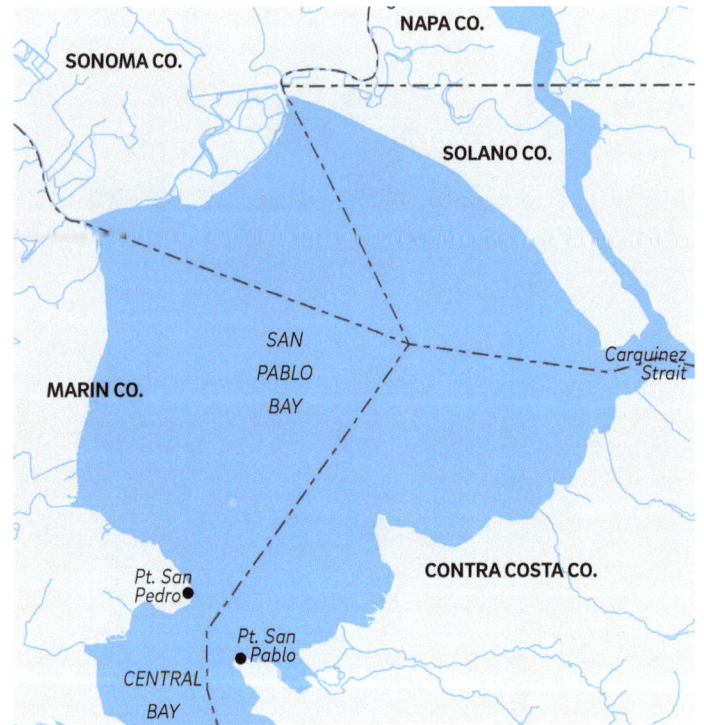

Location map of the study area. This study focuses on San Pablo Bay, in the northern part of the San Francisco Bay **(left)**, and specifically the shoreline between Point San Pedro in Marin County and Point San Pablo in Contra Costa County **(top)**. San Pablo Bay supports many of the tidal marshes in the San Francisco Estuary, and for this reason, it was thought to be a good place to pilot a study of marsh edge dynamics. (imagery courtesy of ESRI)

CONTROLS ON THE MARSH EDGE:
Shoaling and mudflats

To understand the dynamics of the marsh edge, we need to look at the relationship of water depths, wave dynamics, and mudflats to the marsh edge. While the orientation of a given shoreline reach with respect to wind direction often determines its exposure to wave erosion, mudflat width and elevation may be particularly important in determining wave energy reaching the shore. The mudflat serves to temporarily store sediment for resuspension and filter offshore waves. As small waves grow with shoaling, they break or are attenuated due to friction on the mudflat and marsh surface. Wave heights tend to be lower in deep open water, and increase in height close to the shoreline as the water becomes shallower, resulting in higher wave energy at the shoreline (DHI 2011, Veloz et al. 2013).

As waves travel from deep to shallow water, they slow down and steepen due to the decreasing water depth and bottom friction, and break once they reach a limiting depth. At high water levels, such as during storm surges occurring at high tides, waves flood the marsh and attenuate via the same process of depth-limited breaking. In this case, the friction of the vegetation at the surface (together with the mudflat) causes the wave to lose energy.

Within the normal tidal range, mudflats can knock down offshore waves to a lower height; if the mudflat is high enough in the tidal frame, high energy events will only reach the marsh edge at extreme water levels (Lacy and Hoover 2011). Where the mudflat is lower in the tidal frame, or narrow, wave energy at the marsh edge will tend to be higher. Thus, the effects of mudflat slope and shape on shoreline position likely represent a negative feedback loop: the marsh edge may erode, depositing on and widening the mudflat until wave energy is reduced sufficiently so that erosion no longer occurs (Lacy and Hoover 2011). If mudflat elevations do not keep pace with sea level rise, more wave energy will reach the shoreline more frequently, thus increasing exposure of the marsh to higher wave energy and the risk of shoreline erosion (BCDC 2013).

Wave shoaling across a mudflat and marsh. As a wave propagates from deep water to shallow water, the wave length is reduced. The energy flux remains constant and the reduction in speed is compensated by an increase in wave height (and thus wave energy density) which helps explains why wave heights can be higher at the shoreline. However, a wave breaks when it reaches a limiting depth (or when wave height is 0.6 times the water depth) which often occurs over mudflats. (adapted from BCDC 2013)

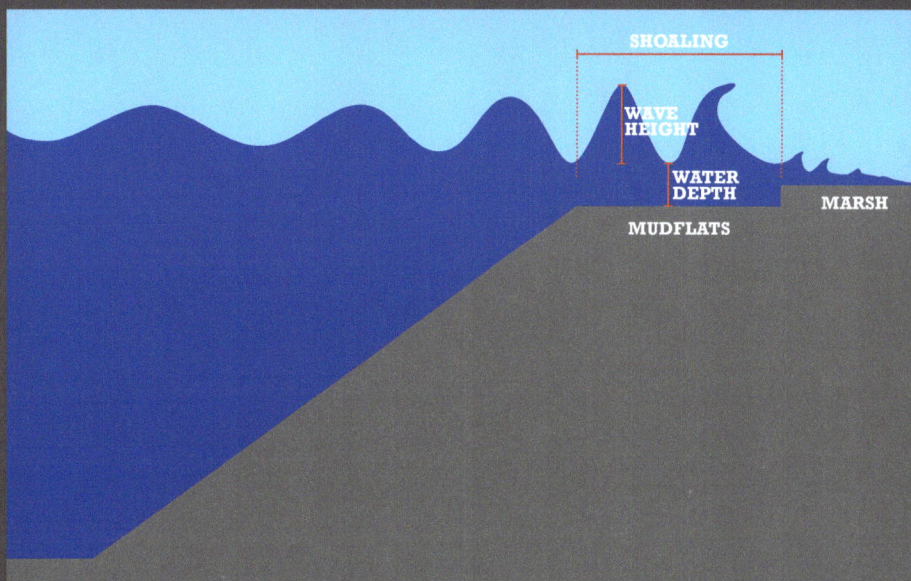

Drivers of change

The major physical drivers controlling marsh edge dynamics in San Pablo Bay include wind wave energy and direction, topography/bathymetry, mudflat elevation, sediment supply, vegetation, relative sea level rise (Allen 1989, Schwimmer 2001, Moller and Spencer 2002, Pedersen and Bartholdy 2007) and other factors such as ferry wakes and biological activity (Pethick 1992, van der Wal and Pye 2004, Francalanci et al. 2011; see below).

WIND WAVE ENERGY AND DIRECTION

Because of San Pablo Bay's proximity to the Golden Gate, wind direction is a significant driving force determining energy directed at the shoreline. Wind direction in San Pablo Bay is mainly northwesterly, with speeds up to 9 m/s during the summer (Jaffe et al. 2007 from Miller 1967). There has been some documentation of a San Pablo Bay Gyre, which rotates clockwise around the North Bay (Walters et al. 1985). The Gyre likely influences wave direction and energy, as well as sedimentation patterns around the Bay. Wave energy in San Pablo Bay tends to be high relative to other sites in the estuary (with significant wave heights of up to 0.6 m) because of several factors, including orientation relative to the prevailing winds and long fetch (Walters et al. 1985, Bever and MacWilliams 2013).

Conceptual model of marsh evolution. This cross section stretches from the subtidal reaches of an idealized shoreline through the marsh to the upland transition zone. It illustrates the different drivers and processes controlling the evolution of the marshes, and of the shoreline in particular. (adapted from PWA)

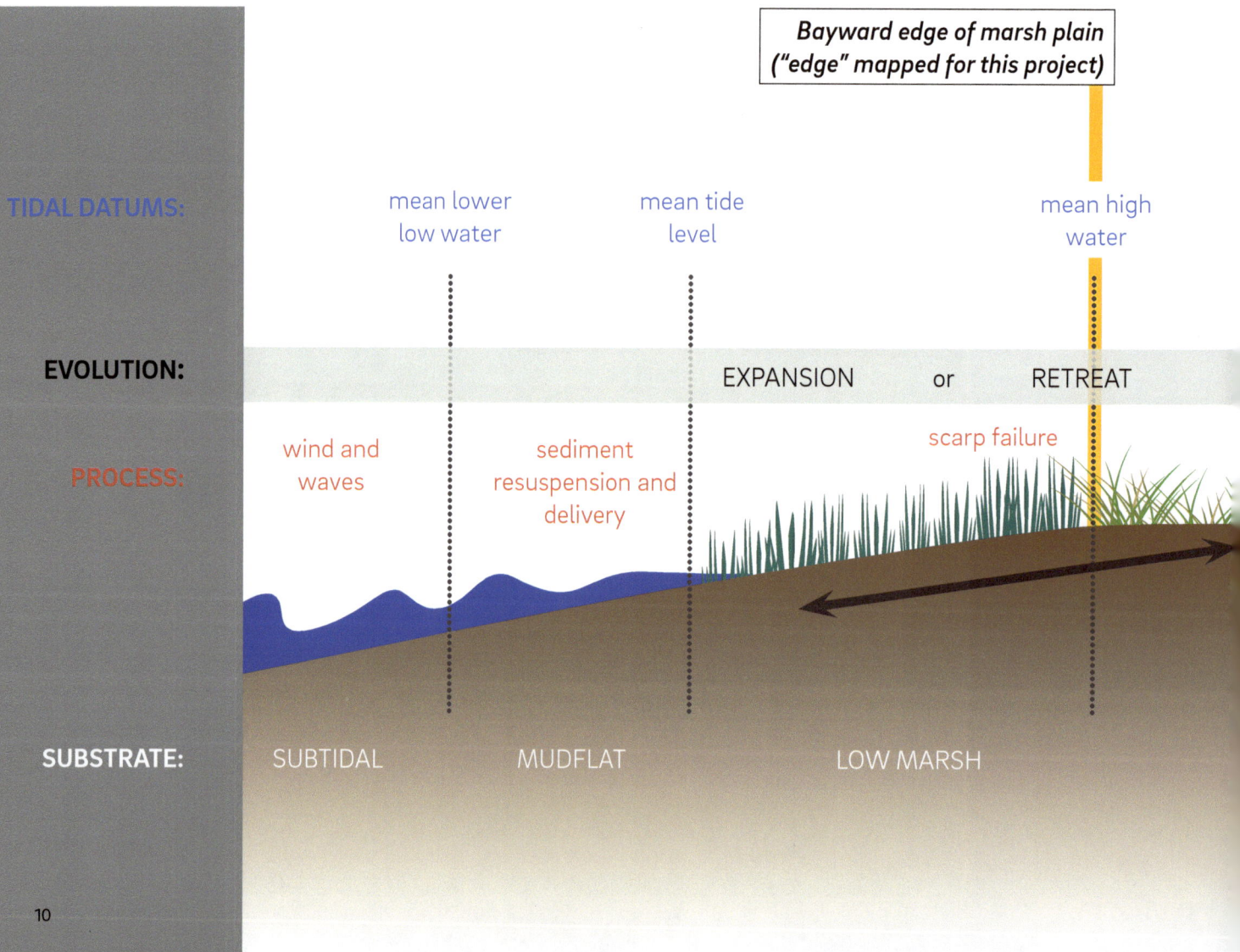

Bayward edge of marsh plain ("edge" mapped for this project)

TIDAL DATUMS:
mean lower low water
mean tide level
mean high water

EVOLUTION:
EXPANSION   or   RETREAT

PROCESS:
wind and waves
sediment resuspension and delivery
scarp failure

SUBSTRATE:
SUBTIDAL
MUDFLAT
LOW MARSH

MUDFLATS

San Pablo Bay is generally shallow (less than 2 m deep at Mean Lower Low Water [MLLW]), with wide mudflats that are exposed at low tides ringing the northern and northeastern sides (Bever and MacWilliams 2013). A deepwater channel extends from the mouth of the Carquinez Strait to the central San Francisco Bay, which averages 12 m in depth (Jaffe et al. 2007).

SEDIMENT SUPPLY CHANGES

San Francisco Bay lies at the bottom of the Sacramento-San Joaquin River watershed which drains approximately 200,000 km² (40% of California). Sediment delivery to the Bay has been dynamic over the last century, and this variability has been expressed as major changes along the shore. During the late 19th century, extensive hydraulic mining in the Sierras coincided with a period of abnormally high regional precipitation, which mobilized large volumes of fine sediment delivered to San Francisco Bay (Gilbert 1917, Barnard et al. 2013). This led to significant changes in bathymetry, as well as the location and extent of beaches and tidal marshes. A comparison of bathymetric surveys in San Pablo Bay between 1856 and 1887 by Jaffe et al. (2007) shows that the estuary accumulated sediments during this period, with intertidal mudflats expanding by 60%.

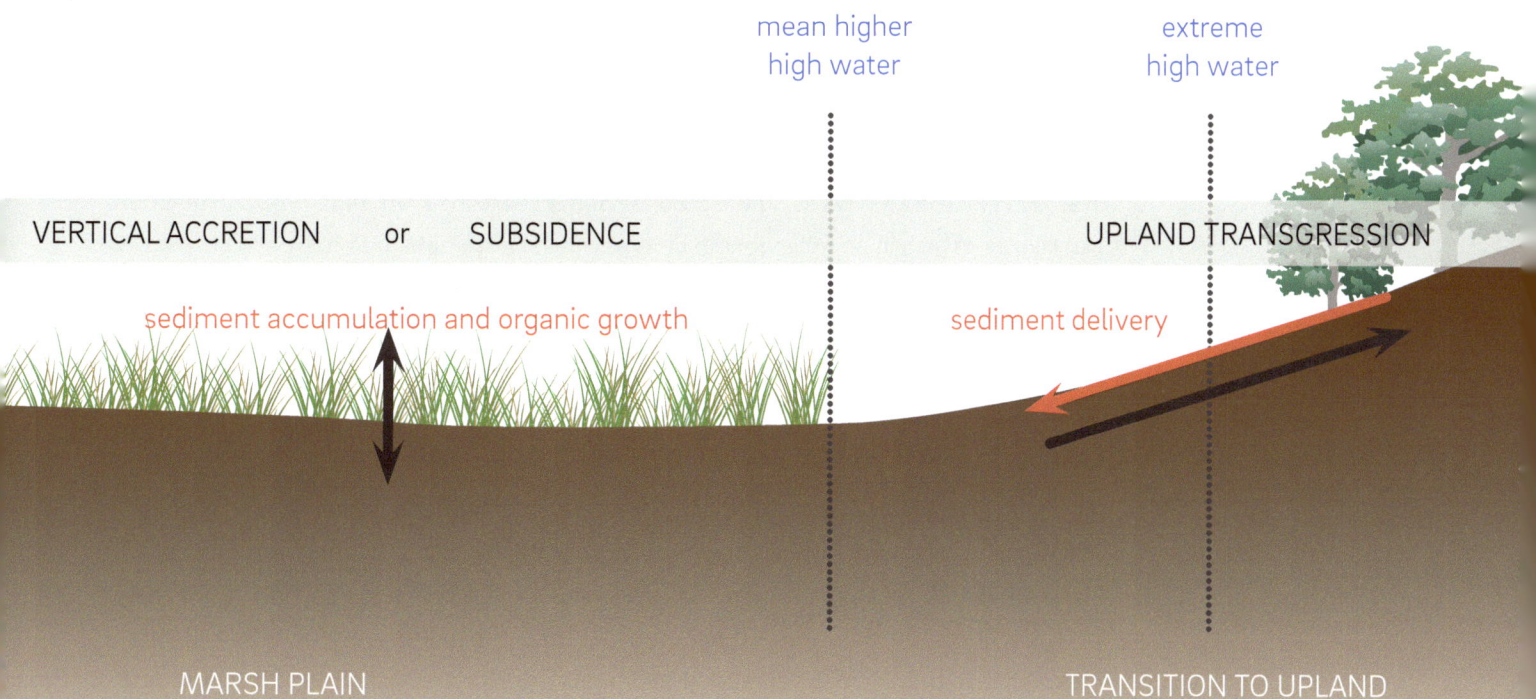

mean higher high water

extreme high water

VERTICAL ACCRETION   or   SUBSIDENCE

UPLAND TRANSGRESSION

sediment accumulation and organic growth

sediment delivery

MARSH PLAIN

TRANSITION TO UPLAND

With this increased sediment delivery, the fringing tidal marshes expanded into the Bay during this period. In 1850, there were 190,000 acres of tidal marsh (Goals Project 1999). In the late 19th century, much of their surface area was reclaimed for farming or development, with dykes and levees holding the shoreline in place.

In the mid-1900s, efforts to manage floods, and develop hydropower and water supply led to the construction of ring dams throughout the Sierra Nevada. The dams, in conjunction with the cessation of mining in 1884, cut off the supply of Sierran coarse sediment to the Estuary (Wright and Schoellhamer 2004, Schoellhamer 2011). Conversely sediment yields from local Bay watersheds increased as a result of levee construction, which isolated flood plains from rivers, logging, urbanization, agriculture and grazing in the mid to late 20th century (Lewicki and McKee 2010).

In the late 20th and early 21st centuries, as local development has slowed and efforts to reduce fine sediment delivery from local watersheds to the estuary have increased, sediment yields have decreased in a number of local watersheds (McKee et al. 2013; see conceptual model below). Reflecting these changes, there has been an observed deficit in suspended sediment concentrations in the San Pablo Bay in recent years (Schoellhamer et al. 2011, Schoellhamer et al. 2013). According to Jaffe et al. (2007), by 1983 the bathymetry of San Pablo Bay had responded to these changes as well, becoming much simpler and net erosional. Most of the side channels filled with sediment and there was widespread erosion on the shallower flats (van der Wegen and Jaffe 2013), leading to an overall loss of mudflats (Goals Project 1999).

VEGETATION PATTERNS

The flux of sediment delivery to the shallows of the estuary is only one part of the story of marsh evolution. Salt tolerant vegetation, such as pickleweed (*Sarcocornia pacifica*) has been shown to be a key factor controlling the evolution of tidal marsh plains and unvegetated tidal channels (Temmerman et al. 2005). Root strength can hold marsh scarps in place, thus increasing the stability of the shoreline (VanEerdt 1985). Vegetation can also re-establish on fallen marsh blocks, which can initiate marsh expansion even in a high energy environment (Allen 1989). The interplay between physical and the biological processes often produces distinct morphologies such as scarps between salt marshes and tidal flats that can influence evolution of the shoreline (Fagherazzi et al. 2012).

RELATIVE SEA LEVEL RISE

As sea level rise accelerates, the depth, duration, and frequency of inundation of tidal marshes could increase (unless sediment supply and bio-accumulation keeps pace), stressing marsh vegetation and resulting in increased wave energy and increased erosive potential along the marsh edge (Fagherazzi 2013). If the nearshore sedimentation rate is higher than the rate of local relative sea level rise, then the marsh edge can prograde (Schwimmer and Pizzuto 2000). If the reverse is true, the mudflat elevation may not keep up with sea level rise, allowing more wave energy to reach the marsh edge.

Previous local studies

Several previous studies have looked at shoreline changes in San Francisco Bay and along the California coast. For this study, we drew on the methods and tools used in earlier studies and adapted them based on our study area and the available data.

Hapke at al. (2006) mapped and measured long- and short-term change along the beaches and bluffs of the California Coast. Using three historical shorelines (1800s, 1920s-1930s, and 1950s-1970s) and a recent shoreline derived from LiDAR, they calculated rates of change for 45% of the California coastline.

Zoulas (2006) estimated long-term shoreline change rates in the vicinity of Corte Madera Creek in Marin County as part of a broader effort to understand historical sediment trends in the adjacent tidal marsh. Shoreline change rates from 1853 to 2006 were derived from 11 historical shoreline positions, which were digitized from georeferenced aerial photographs, U.S. Coast Survey T-Sheets, and USGS Digital Orthophoto Quadrangles (DOQs). The entire study area experienced substantial erosion from 1853 to 2006; the degree of inland migration averaged 148 m over this time period, though the erosion was not spatially or temporally uniform.

Doane (1999) documented variable rates of erosion and progradation at several sites in San Pablo Bay using historical maps and field data, and identified marsh scarps in certain locations. Doane found an overwhelming trend towards erosion between 1951 and 1997, though the shoreline was highly variable over this period, with short-term cycles of erosion and progradation.

3. METHODS

Defining the marsh edge

Challenges

Mapping the transition from land to water (the shoreline) poses a challenge because, despite its name, the shoreline is not a fixed line in the landscape. Within the intertidal zone, there are several lines one might consider designating as "the shoreline," including mean tide level (MTL) or mean lower low water (MLLW). The intertidal zone can also be very dynamic over space and time. Mapping the shoreline using remote data sources poses the additional challenge of determining which features on the landscape correspond with the tidal data from that time period.

How the edge was identified

For the purposes of this study, we aimed to map a persistent feature, in order to consistently map and calculate changes over time. We defined the shoreline here as a geomorphic transition between mid-marsh and low marsh, or the bayward edge of the marsh plain. This edge separates the unconsolidated, low marsh dominated by cordgrass (Spartina spp.) and mostly unvegetated mudflats, from the landward, consolidated, mostly vegetated mid-marsh dominated by pickleweed (Sarcocornia pacifica; see pages 10-11).

A visible transition between low marsh and mid-marsh was identified in the field and matched with the corresponding signature in aerial imagery, oblique images, and landscape photography (right). We used this definition to guide our GIS mapping procedures for delineating this feature at three points in time. This marsh edge transition took a variety of forms, as shown in the next two pages.

The mapped shoreline. The red-dotted line in this field photo **(top)** and aerial photo **(below)** corresponds to the boundary between the consolidated mid-marsh and unconsolidated cordgrass-dominated emergent low-marsh (where staff sunk up to their knees). Field photo and aerial imagery are from the same location along the Highway 37 marsh in Solano County. ([top] photo by Julie Beagle, April 2012; [bottom] NAIP 2010)

MARSH EDGE TYPES

Five distinct marsh edge types were identified in the field and confirmed with LiDAR and aerial photos (right). These types were used to determine where the shoreline should be digitized. The typology was based on the presence or absence of scarps, the presence or absence of vegetation, and the inflection (rapid flattening) of the slope. The five edge types include 1) scarps with bayward vegetation (SV), 2) scarps without bayward vegetation (SN), 3) ramps with inflection points (RI), 4) ramps without inflection points (RNI), and 5) beaches fronting marshes (B).

Scarps (SV and SN), generally less than two meters high, were identified along parts of the shoreline using oblique imagery (Google Earth, BING maps, Salomon 2012). Where scarps were present, the shoreline was digitized at the bottom of the scarp to include the bare earth or exposed face of the scarp. Where some vegetation was visible past the bottom of the scarp extending towards the Bay (SV), the shoreline was still digitized at the bottom of the scarp. The digitized shoreline separated the bottom of the scarp on the "land side" from the presumed ephemeral or emergent vegetation on the "bay side."

Where no scarp was discernible in the oblique imagery, and the profile of the shoreline was more like a ramp, we looked for an inflection in slope. The shoreline was digitized along the inflection in slope (RI). If no inflection point was visible, we digitized the shoreline at a visible transition in vegetation signature (RNI). If a wrack line indicating a single-ridge marsh berm was visible, we always digitized the shoreline bayward of the marsh berm at the transition of the vegetation signature. In the Bing oblique imagery, vegetation with a brighter green signature was presumed to be younger, and the shoreline was digitized with this brighter vegetation on the bay side.

Beaches and rocky shorelines were digitized at the transition between marsh and beach.

Similar marsh edge types were identified in England by Moller and Spencer (2002) and Allen (1989). Prahalad et al. (2015) described slopes in Tasmania that vary from gently sloped grassy ramps to near-vertical and even overhanging clifflets that expose sediment.

Shoreline types in profile with aerial photos (right). Oblique imagery is paired with cross sections abstracted from a 2010 LiDAR DEM demonstrate the different profiles of the marsh edge. ([left] photo by Shira Bezalel, February 2015; [right] photos by Micha Salomon, February and April 2012)

1 Scarp with bayward vegetation (SV)

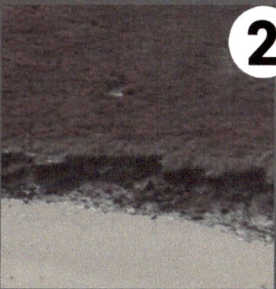

2 Scarp without bayward vegetation (SN)

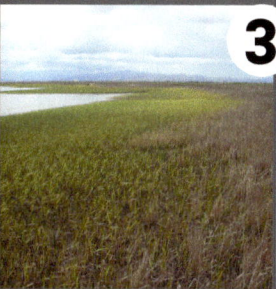

3 Ramp without inflection point (RNI)

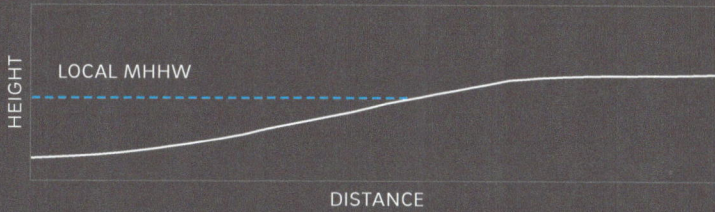

4 Ramp with inflection point (RI)

5 Beaches or rocky shoreline (B)

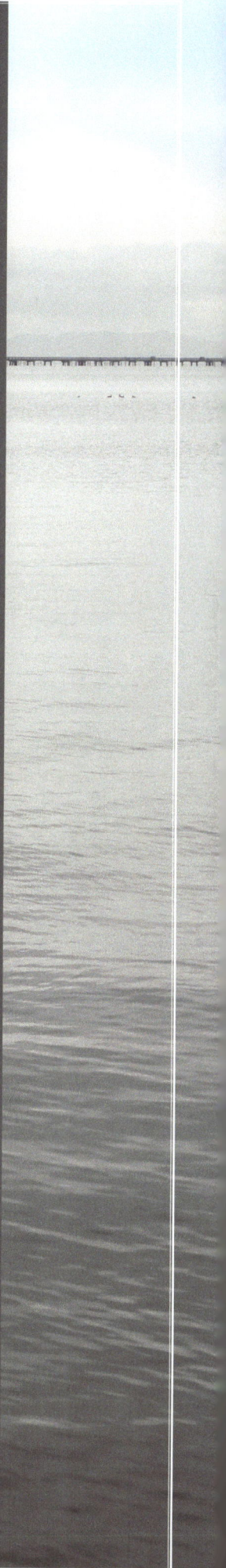

Mapping the shoreline

Overall approach

We mapped the San Pablo Bay shoreline at three points in time: ca. 1855 (the historical shoreline), 1993 (the recent shoreline), and 2010 (the contemporary shoreline). We were thus able to measure both long-term (1855-1993) and short-term (1993-2010) changes in shoreline location (right). In order to ensure consistency in feature mapping through time, we developed protocols for mapping the shoreline for each of the time periods. Only areas of shoreline with marshes present (or beaches with marsh behind them) were mapped. All of the shorelines were mapped in North American Datum 1983 (NAD83), UTM Zone 10 North.

Mapping the 2010 shoreline

To map the contemporary shoreline, we used a combination of data sources, including 2010 NAIP imagery, 2010 LiDAR-derived DEMs and Hillshades (NOAA/OPC), vertical and oblique imagery available from Google Earth and BING maps, low altitude oblique aerial photography, and landscape photography from field visits. The shoreline was digitized from aerial photography at a scale of 1:1,000. Visible crenulations in the shoreline on the order of a few meters in size were digitized, to the extent that they were identifiable in the 1-meter spatial resolution source imagery.

A decision tree was developed to help determine where the shoreline should be mapped in different situations, and to ensure consistency across time periods as well as limit human error in digitizing (see next page). The factors taken into account in the decision tree included shoreline edge type, microtopography (at the meter scale), and vegetation patterns and transition. In many cases vegetation type was used as a proxy for topography. Oblique imagery and the LiDAR DEM was used to identify subtle changes in slope, especially inflection points where steep slopes suddenly transitioned into flats, and the relief of tidal wave-built berms.

Examples of sources used to map the shoreline. The primary data sources used to map the three shorelines were the T-sheets from the late 19th century **(top right)**, the 1993 digital orthoquads **(middle right)**, and the 2010 NAIP imagery **(bottom right)**. Ancillary data was used in all three mapping efforts to increase certainty in the location of the shoreline. (Lawson and Welker 1887; courtesy of NOAA, DOQQ 1993, NAIP 2010; [below] photo by Ruth Askevold, April 2015)

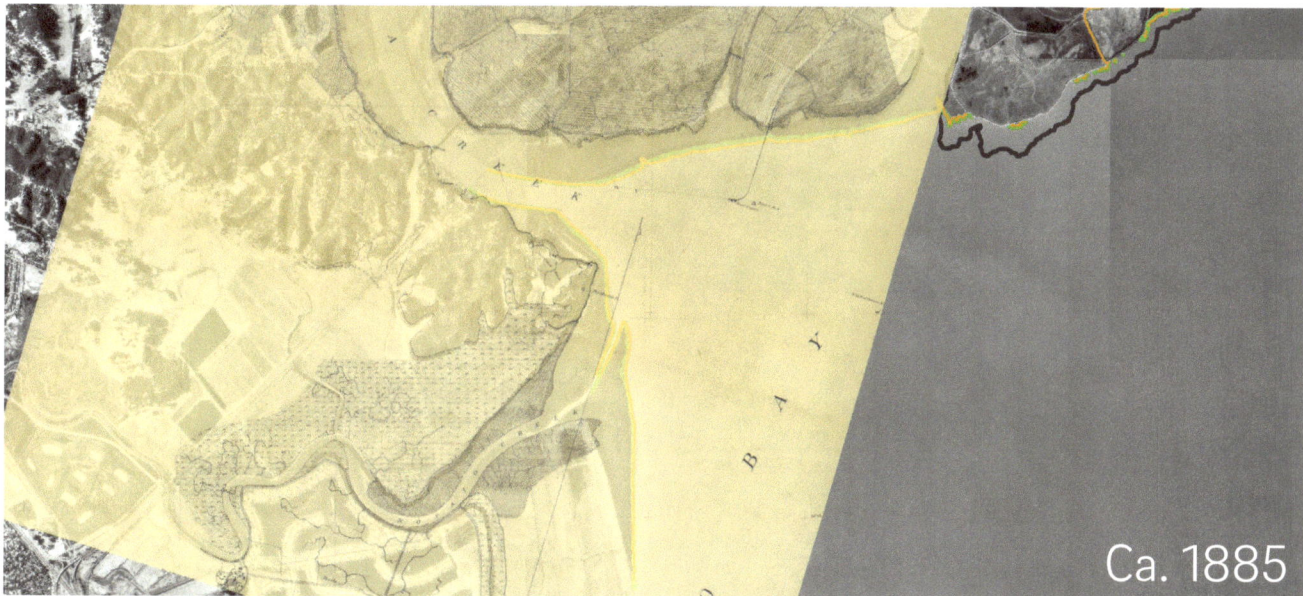

Ca. 1885

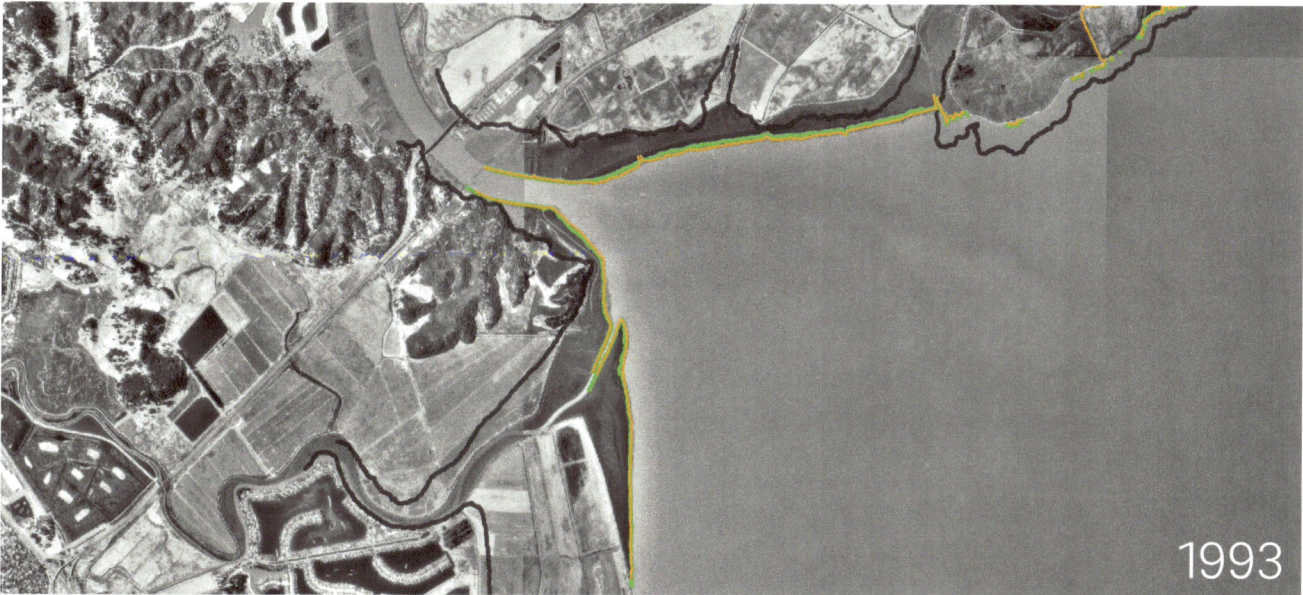

1993

2010

Contemporary (2010) decision tree

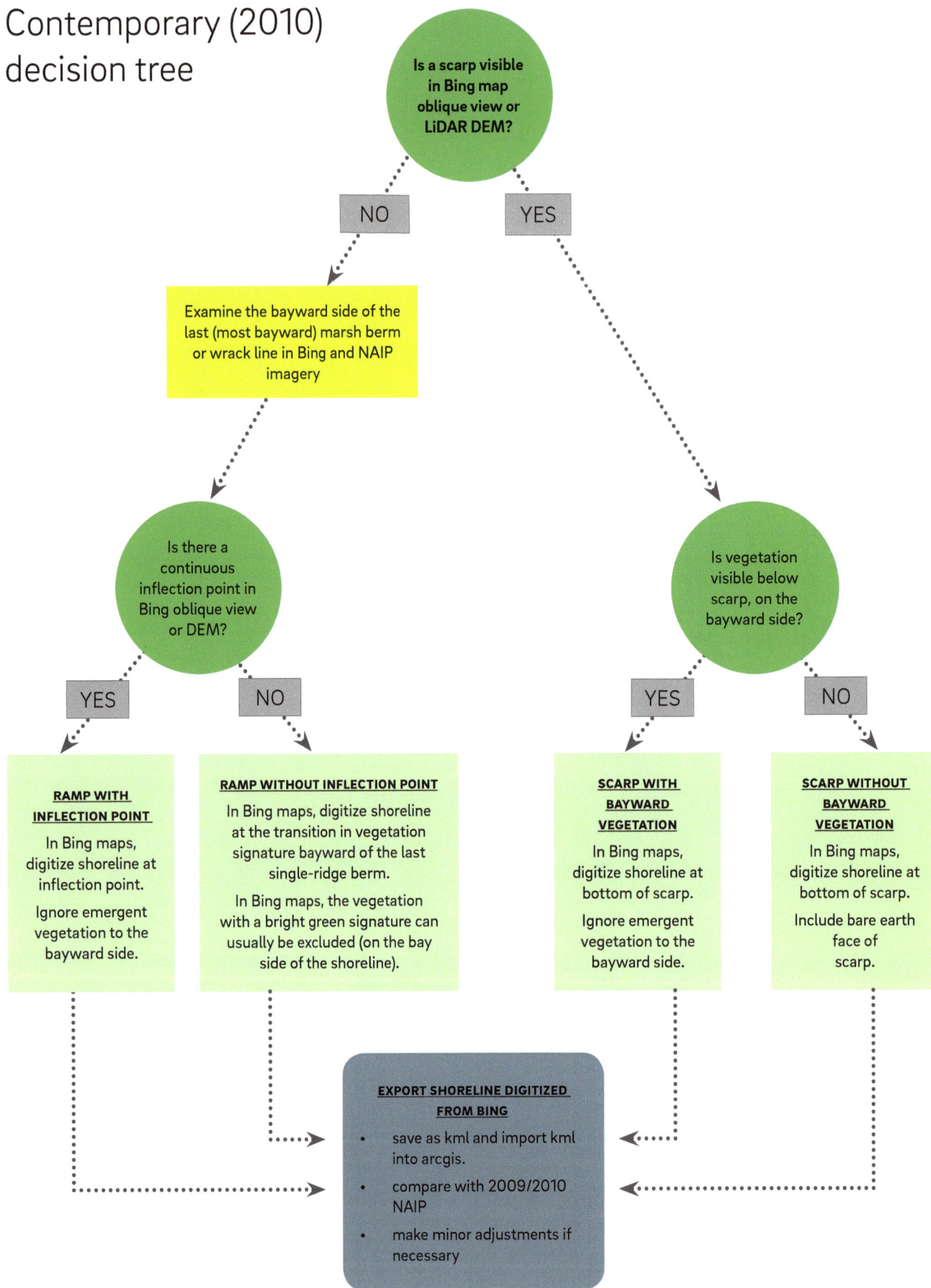

Is a scarp visible in Bing map oblique view or LiDAR DEM?

NO →

Examine the bayward side of the last (most bayward) marsh berm or wrack line in Bing and NAIP imagery

Is there a continuous inflection point in Bing oblique view or DEM?

YES →

RAMP WITH INFLECTION POINT

In Bing maps, digitize shoreline at inflection point.

Ignore emergent vegetation to the bayward side.

NO →

RAMP WITHOUT INFLECTION POINT

In Bing maps, digitize shoreline at the transition in vegetation signature bayward of the last single-ridge berm.

In Bing maps, the vegetation with a bright green signature can usually be excluded (on the bay side of the shoreline).

YES →

Is vegetation visible below scarp, on the bayward side?

YES →

SCARP WITH BAYWARD VEGETATION

In Bing maps, digitize shoreline at bottom of scarp.

Ignore emergent vegetation to the bayward side.

NO →

SCARP WITHOUT BAYWARD VEGETATION

In Bing maps, digitize shoreline at bottom of scarp.

Include bare earth face of scarp.

EXPORT SHORELINE DIGITIZED FROM BING

- save as kml and import kml into arcgis.
- compare with 2009/2010 NAIP
- make minor adjustments if necessary

At any given point, the shape and detail of the digitized shoreline is dependent upon the quality of the source imagery in that particular location. The timing of the source imagery is important: in photos taken during high tide, subtle shoreline features can be covered by water and difficult or impossible to discern.

Acquisition of NAIP imagery is not tidally controlled, although indices of individual NAIP tiles contain the acquisition date of the imagery and imagery was acquired within about two hours before or after noon to minimize shadows. Thus, by consulting historical tide charts, and by comparing the relative extent of mudflat and open water in the primary imagery with other aerial images of the same location, we were generally able to determine whether the imagery was taken during high, low, or mid-tide. In a few locations, 2009 data was used as a digitizing source for the contemporary shoreline because it was taken at a lower tide.

When available, additional imagery was used to help interpret or confirm the identity of visible features in the digitizing sources. Most of the NAIP 2005 imagery was taken at low tide, making for useful comparisons with the less consistent (in terms of tidal cycles) 2009 and 2010 imagery. The dates of the aerial imagery used for mapping were recorded for each shoreline segment in the GIS database. This date information was also used in the analysis to determine rates of shoreline movement over time.

The presence of tidal creeks and rivers complicated shoreline mapping somewhat. A set of rules based on channel size was devised to determine how far along the banks of tidal channels shoreline mapping would continue. Small channels with mouths less than 3 m wide were ignored; the shoreline was digitized straight across these smallest channels. For small and medium-sized channels (3 m to 25 m wide at the mouth), the shoreline was not digitized and gaps were maintained where the channels entered the Bay. Major channels, including all named rivers and creeks, were digitized for some distance up the channel, along the banks. Shoreline digitization stopped wherever one of these three criteria were met: the channel passed under a bridge, the channel had a major split into two or more large channels, or the digitizing source became blurred and difficult to interpret.

Decision tree developed to map the 2010 shoreline. This decision tree was developed so that the process of mapping the 2010 shoreline could be transparent and repeatable. It also allowed the shoreline to be classified by the type of edge morphology observed. The primary sources of data were the NAIP imagery, Bing oblique imagery and the 2010 LiDAR dataset. This decision tree was then used to QAQC the dataset.

Mapping the 1993 shoreline

The 1993 shoreline mapping process used a similar approach to the contemporary mapping, but posed different challenges. Some of the differences and challenges are detailed here.

We used DOQQ black and white high resolution images to digitize the 1993 shoreline. This imagery is comparable to the 2010 NAIP imagery, but there are some important differences. NAIP imagery contains three color bands and has a 1 m spatial resolution, while the 1993 imagery (DOQQ) is greyscale and has lower spatial resolution, which often made it difficult to discern subtle differences in vegetation signatures. There was also an absence of regionally consistent ancillary data, such as oblique or landscape photography, for this time period.

A simple decision tree was developed for the 1993 mapping which takes into account the presence or absence of a white wrack line in the imagery, often deposited against a single-ridge marsh berm (right). Where one or more was present, the shoreline was digitized bayward of visible wrack lines. If no wrack line was present, the limit of a dark marsh signature was delineated. In both cases, the dark marsh signature was digitized, but we assigned higher confidence values to segments which were mapped using a wrack line. For some parts of the shoreline, 1995-96 NASA imagery was used to aid in the interpretation of the DOQQ imagery.

Decision tree for mapping the 1993 shoreline (right). A simple decision was used to map the 1993 shoreline. This depended on the identification of a white line, usually indicating a wrack line, on the front side of a marsh berm. If one was visible, this greatly aided the certainty in assigning a shoreline location. When it was not available, or a different process of shoreline evolution was in effect, we used the boundary of a dark marsh signature. Ancillary data from 1995-96 was used in some locations.

Examples of 1993 shoreline mapping with uncertainty (below, right and far right). Two examples of the 1993 DOQQ highlight some of the challenges with mapping the marsh edge. In the example on the left, the dark marsh signature is difficult to distinguish from the Bay, resulting in a low certainty value. In the example on the right, several wrack lines and a consistent change in the greytones make finding the shoreline a bit easier. **(below)** 1993 DOQQ shown at a scale of 1:4,000 showing a wrack line in white and a fuzzy marsh edge location.

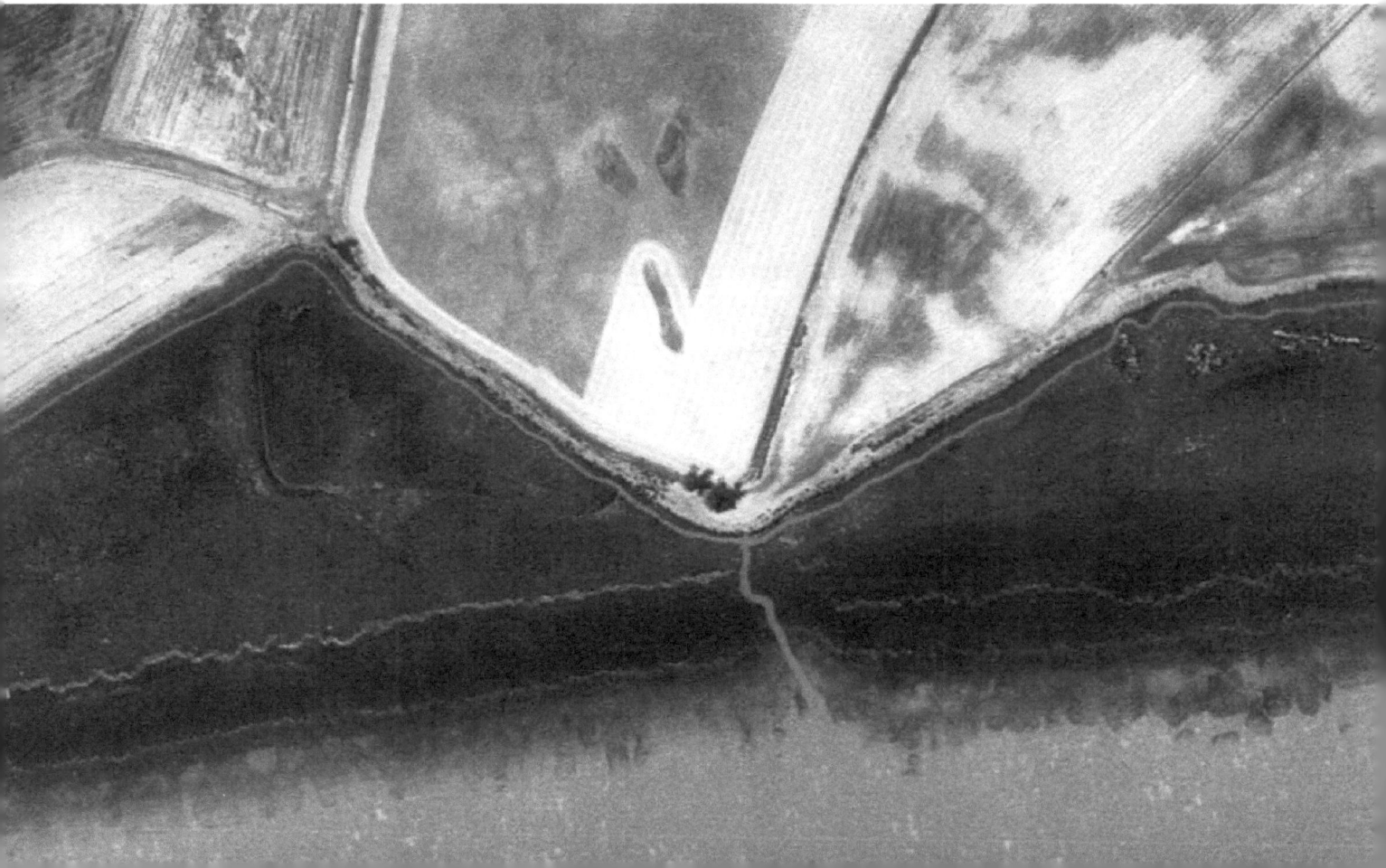

1993 decision tree

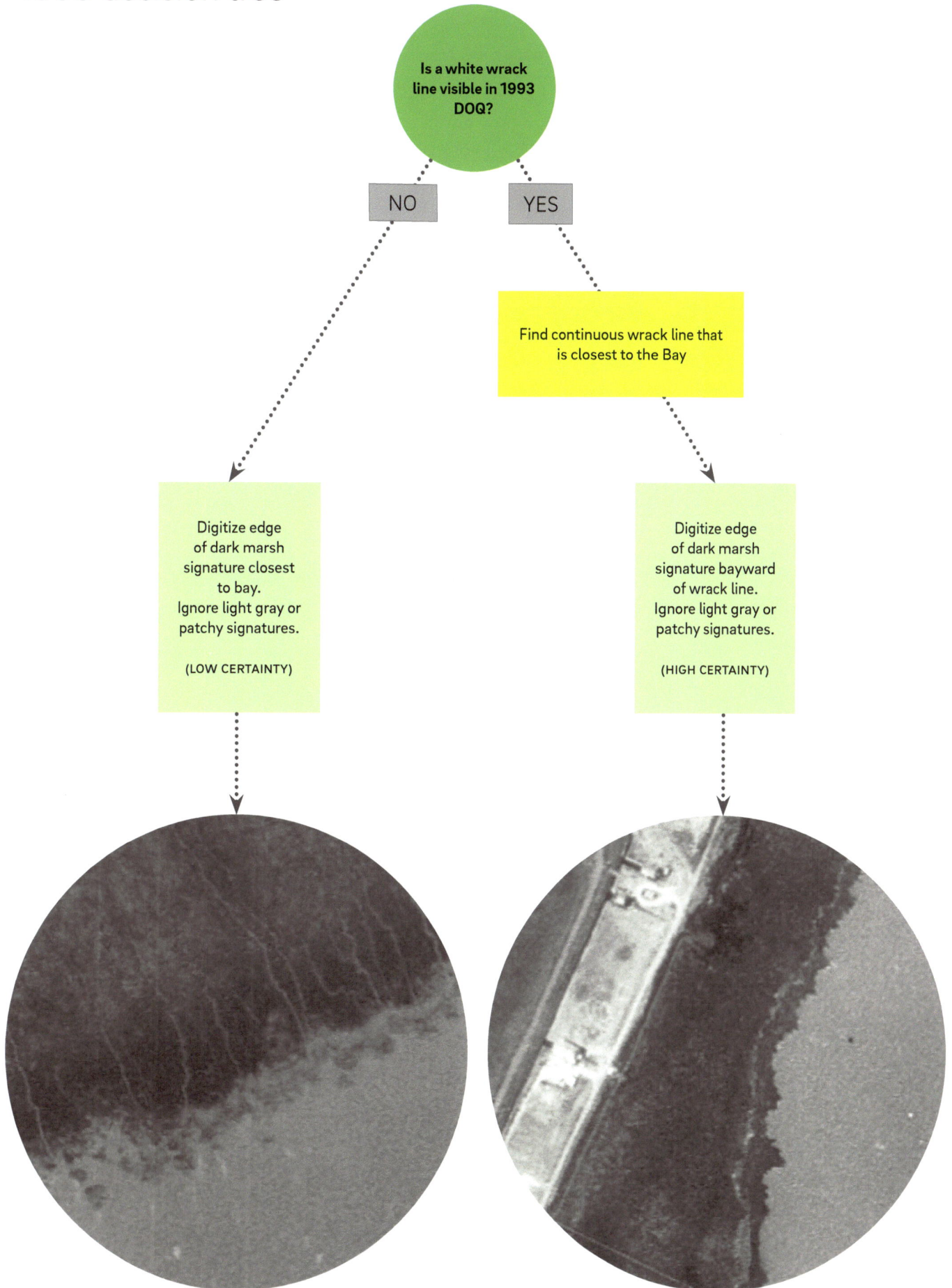

Is a white wrack line visible in 1993 DOQ?

NO

YES

Find continuous wrack line that is closest to the Bay

Digitize edge of dark marsh signature closest to bay.
Ignore light gray or patchy signatures.

(LOW CERTAINTY)

Digitize edge of dark marsh signature bayward of wrack line.
Ignore light gray or patchy signatures.

(HIGH CERTAINTY)

Mapping the ca. 1855 shoreline

The ca. 1855 shoreline was derived from the Historical Baylands GIS layer from SFEI EcoAtlas 1998 (available at: http://www.sfei.org/sites/default/files/ EcoAtlas_SFEI.zip), which is in turn based on the shoreline depiction in U.S. Coast Survey (USCS) T-sheets. The USCS was established in 1807 to create navigation maps of the coastline and immediately adjacent areas. The maps covering the landward portion of the coastline, known as "topographic sheets" or "T-sheets," are a highly valuable source because of their large scale, remarkable detail, and high scientific standards. For examples of annotated T-sheets and discussion of T-sheet symbols, see Grossinger et al. 2011 and Shalowitz 1964.

In general, the ca. 1855 shoreline corresponds to the boundary between vegetated tidal marsh and and unvegetated tidal flat as depicted in the T-sheets. Since the marshes of San Pablo Bay are for the most part not inundated at mean high water, this definition of the shoreline is not in conflict with the delineation method used for the 1993 and 2010 shorelines. In a few locations where no tidal flat was mapped by USCS surveyors, the shoreline was defined as the boundary between beach and shallow bay.

T-sheet depicting the mouth of Petaluma River. This map was created in 1856 and shows the marsh edge from the mouth of the Petaluma River and eastward. The edge of the marsh plain is drawn with a thin black line, separating the partially vegetated low marsh (hashed lines) from the open bay. By this time, levees had already been build around the marshes, beginning the reclamation process, and fringing marsh had begun to develop on the bayward side. (Lawson and Welker 1887; courtesy of NOAA)

Measuring shoreline change

Once the 2010, 1993, and ca. 1855 shorelines were digitized, we used a publically available GIS tool to calculate shoreline change metrics for two time intervals (1855-1993 and 1993-2010). The Digital Shoreline Assessment System (DSAS) is a free software program developed by the USGS that computes rate-of-change statistics from multiple historic shoreline positions residing in a GIS database, and incorporates error and uncertainty into its outputs (Thieler et al. 2009). The metrics we calculated included net shoreline movement (NSM) in meters, and annual endpoint rate (net shoreline movement per year; EPR) in meters/year.

Using DSAS requires following several steps. First, a baseline is created that is roughly parallel to the digitized shorelines. Second, perpendicular transects are cast from the baseline to intersect with the digitized shorelines. Lastly, the points of intersection between the transects and shorelines are used to calculate metrics of shoreline change (below).

Baselines were created using a modified buffer. A 15 m linear buffer was created for the ca. 1855 shoreline and for a combination of the 1993 and 2010 shorelines. The buffer was manually edited near all endpoints to reduce the generation of errant transects. A combination of onshore and offshore baselines were used for both time periods, and as a general rule baselines were created so that transects would diverge rather than converge, if the convex side of the baseline was nearest to the shorelines in question.

Transects were spaced 20 m apart (this interval was selected after running a sensitivity analysis to test different spacing intervals). In DSAS, the transects were set to cast perpendicular to a 'smoothed' baseline. Smoothing values were set at 1000 m for the Sonoma and Solano county shorelines and at 300 m for the Contra Costa and Marin county shorelines. After NSM and EPR were calculated, the transects were clipped to the shoreline change envelope to aid in visualization.

Schematic of shoreline erosion measurement (below). This figure shows the location of an idealized shoreline at two points in time (red dots and purple dashes), and the transects (dotted lines) extending from the baseline (brown line). Areas of marsh expansion are highlighted in blue; areas of erosion in green, and no change is noted by a yellow circle.

- • • • • • 2010 shoreline
- – – – 1993 shorelline
- ▮ progradation
- ▮ erosion
- ● no change
- ▬ baseline
- • • • • • • transects

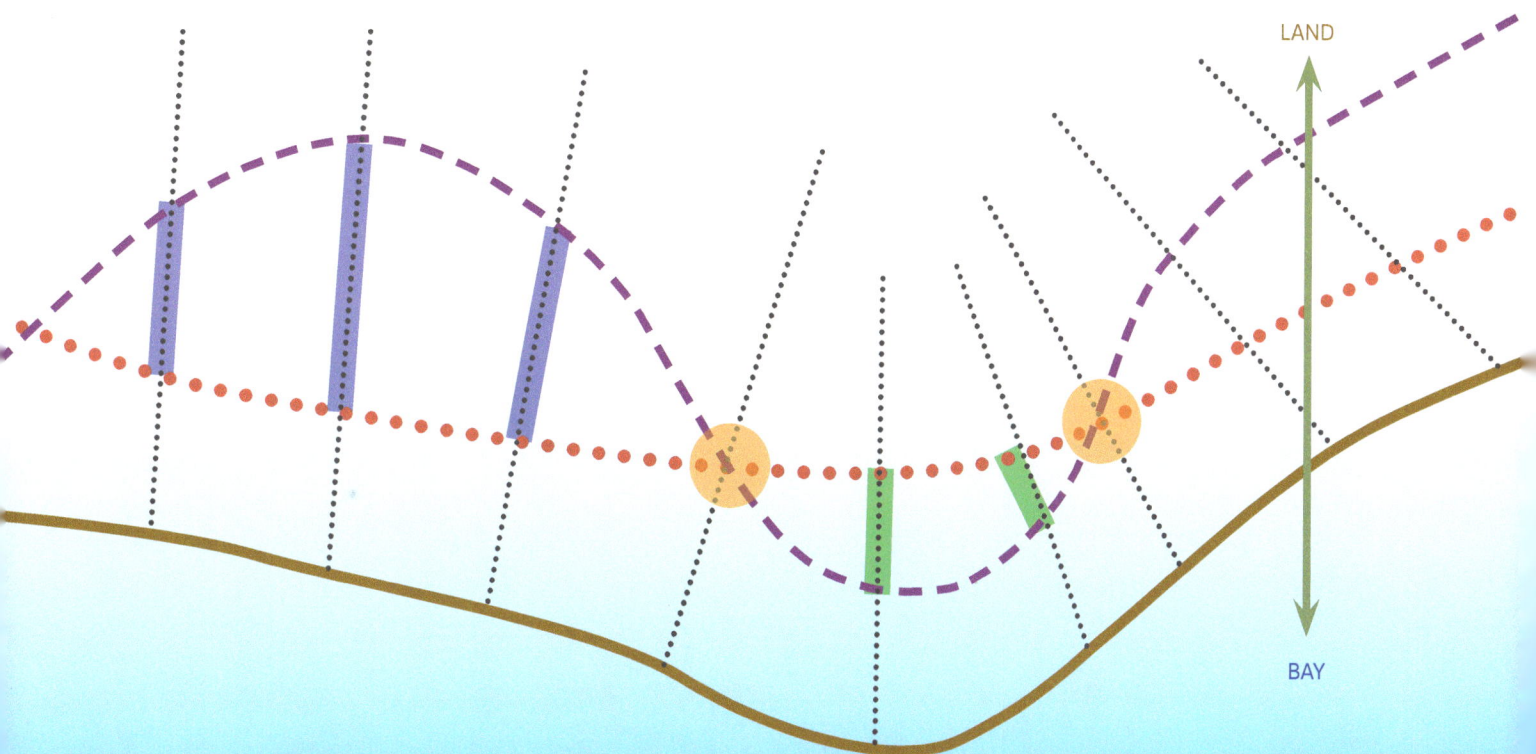

LAND

BAY

Mapping uncertainty and error estimation

We estimated uncertainty in shoreline location for each of the three time periods based on the data sources used to map the shorelines. Mapping uncertainty arose from several sources. For the 1993 and 2010 shorelines, the transition between low marsh and mid-marsh was not equally visible in all areas of San Pablo Bay. Visibility depended on shoreline edge type, image clarity, tide level, and other factors, and thus different parts of the shoreline had different amounts of positional uncertainty. For the ca. 1855 shoreline, uncertainty arose from the fact that the T-sheets for this study were georeferenced without the aid of a modern GIS.

Certainty values for each shoreline by percentage of total length (right). This graph shows the breakdown of certainty values for each shoreline. The ca. 1855 shoreline was mapped with low certainty (meaning the location of the shoreline could be off by as much as 50 m in either direction). Ninety-nine percents of the 1993 shoreline was mapped with medium certainty (±10 m in either direction). More than 60% of the contemporary shoreline was thought to be mapped with high certainty, and the rest with medium certainty.

Pt. Pinole shoreline, looking south (below). (imagery courtesy Google Earth)

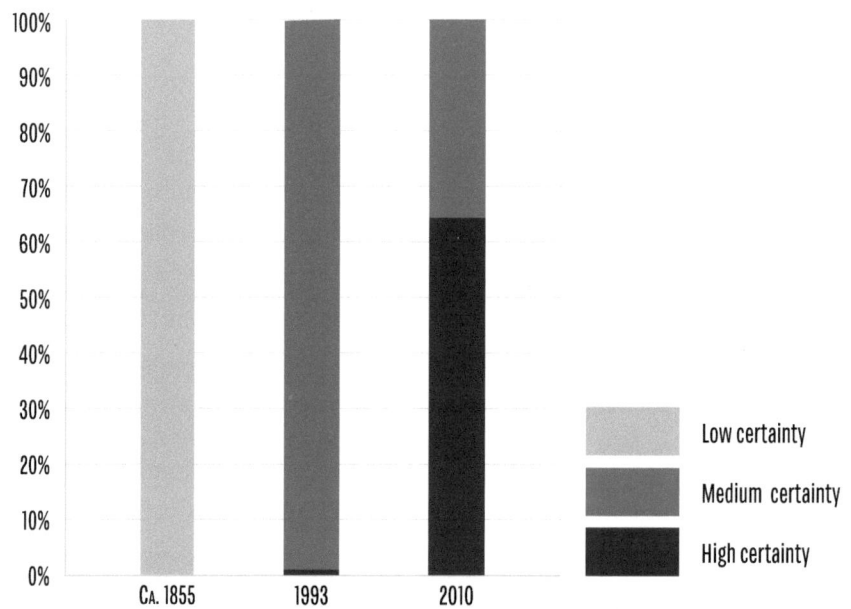

Legend:
- Low certainty
- Medium certainty
- High certainty

In order to account for differing degrees of uncertainty, each segment of the mapped shoreline was assigned an uncertainty value of High, Medium, Low, or T-Sheet. Uncertainty values were estimated in meters based on the interpretability of the imagery at that location. Values of 2, 10, or 20 meters were assigned to segments of the 1993 and 2010 shorelines based on the interpretability of the digitizing imagery and ancillary data (Table 1). An uncertainty value of 50 meters was assigned for the entire ca. 1855 shoreline to account for various sources of error. A single mid-year date of July 1, 1855 was assigned for this shoreline.

The DSAS tool incorporates the error assumed in the creation of the shorelines with the error estimated from many other cumulative sources required to produce this dataset. In a 2011 study along the New England Mid-Atlantic shoreline, Hapke et al. (2011) identify several areas of uncertainty, the most applicable of which we considered and adapted in this study. For more detail on uncertainty, please see Appendix A.

Table 1. Descriptions of uncertainty values for mapped shorelines. Different certainty values were assigned to different shoreline lengths depending on the certainty of the person digitizing, and the source data.

Class	Uncertainty Value	Years	Description
High	2 m	1993, 2010	shoreline is clear and confirmed by other sources
Medium	10 m	1993, 2010	shoreline is somewhat clear in at least 1 source
Low: 1993, 2010	20 m	NA	shoreline is blurred
Low: T-sheet	50 m	ca. 1855	T-sheet HWL

4. RESULTS

In this section, we report the general trends of lateral shoreline change around San Pablo Bay, identifying which areas of marsh have expanded, retreated, or remained relatively stable. We then dive deeper into the short term results (1993-2010) and interpret patterns of marsh edge change as indicated by differences in setting and morphology.

Location of the marsh edge over time

The results of our shoreline mapping are shown in here, with three zoom-ins to highlight some of the different settings in which the marsh edge was mapped over time. The pink line represents the location of the historical marsh edge ca. 1855, the green line represents 1993, and the yellow represents 2010. In the next two spreads, we show the differences between the two time periods, to highlight rates of change, patterns, and management implications of these findings.

Results of mapping three shorelines of San Pablo Bay (left). This map shows the locations of the ca. 1855 (pink), 1993 (green), and 2010 (yellow) shorelines in San Pablo Bay, focusing on the marsh edge. In the ca. 1855 shoreline some locations of rocky shorelines are mapped. (imagery courtesy ESRI)

— ca. 1855 shoreline

— 1993 shoreline

— 2010 shoreline

N

1:100,000

2 miles

Sonoma
Creek

Petaluma
River

Tolay Creek

Novato

SAN

PABLO

BAY

HAMILTON
FIELD

Gallinas Creek

POINT
PINOLE

San Rafael

San Pablo
Creek

N

1:100,000

1 mile

Richmond

Napa River

Vallejo

MARE ISLAND

Carquinez Strait

Pinole

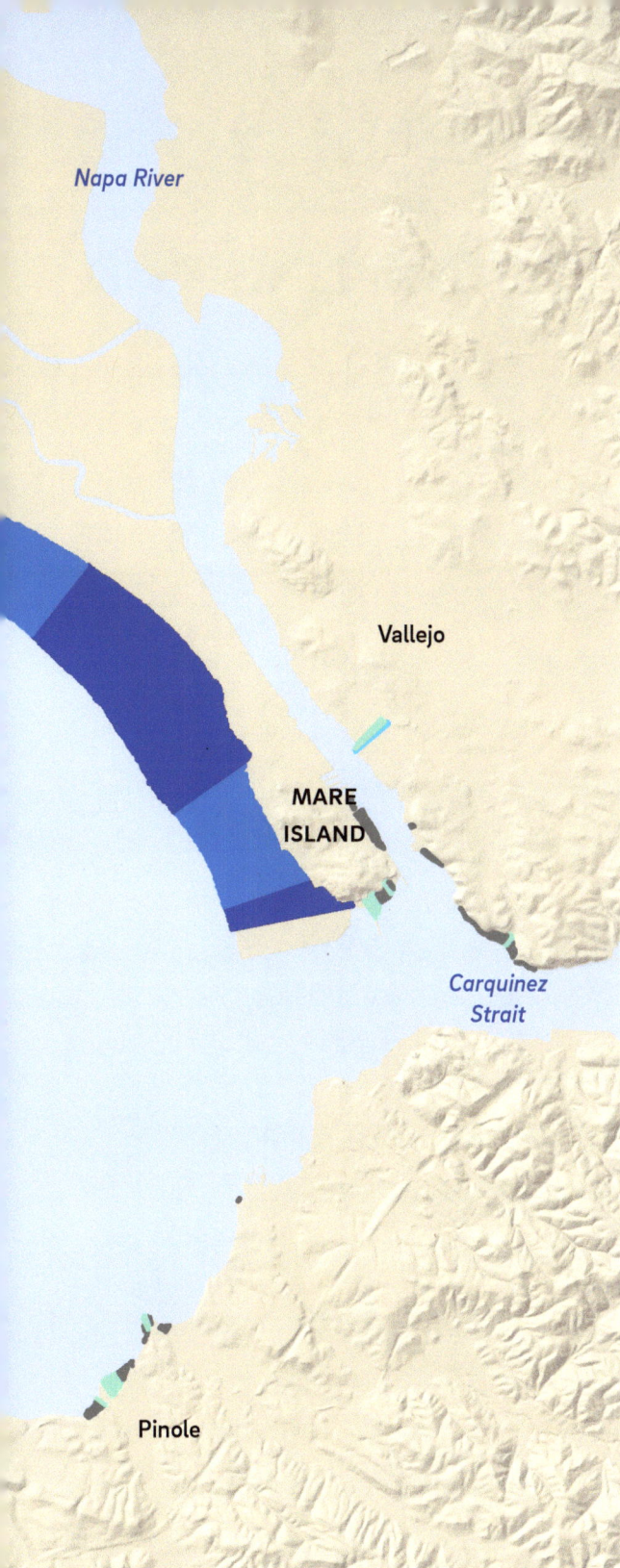

Hydraulic mining during the Gold Rush caused large pulses of sediment to be delivered to San Pablo Bay. As the rate of basin infilling outpaced sea level rise and the erosional pressure of waves, vertical accretion and outward expansion resulted in growth of marsh area and a dramatic change in the San Pablo Bay shoreline (Gilbert 1917, Atwater et al. 1979, Schwimmer and Pizzuto 2000, Fagherazzi et al. 2006). Overall, 62% of the San Pablo Bay shoreline was found to have advanced between ca. 1855-1993. Marshes southwest of Mare Island and on the west side of the Bay expanded by as much as 1600 m into the Bay. This period also saw rapid population growth and development of local watersheds, resulting in increased local sediment supply to the Bay (McKee et al. 2006). The creek deltas of Gallinas Creek, Sonoma Creek, Petaluma River, and San Pablo Creek prograded by as much as 1-5 m/yr between ca. 1855 and 1993. At the same time, widespread reclamation of the marshlands cut off sediment delivery to existing marshes and levees tried to hold the shoreline in place (Dedrick and Chu 1993). Within this overall trend of marsh expansion (and reclamation), modest erosion (on the order of 1-3 m/yr) was documented on headlands such as Point Pinole and the protrusion near the mouth of Tolay Creek. Less than 2% of the mapped shoreline was found to have eroded over this time period. It should be noted that much of the change in this time period took place in the decades around the turn of the 20th century, so rates were even higher at times (and often relatively stable in the latter half of the 20th century).

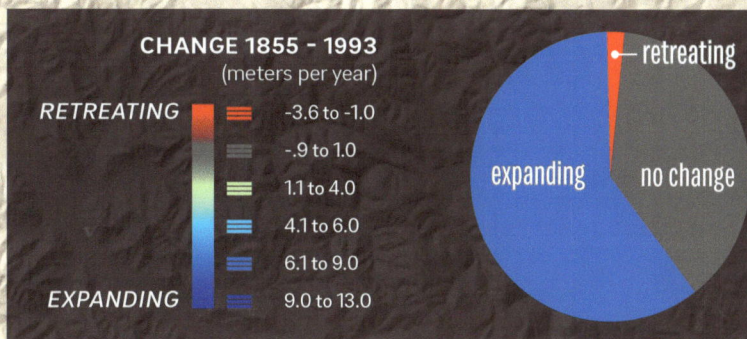

CHANGE 1855 - 1993
(meters per year)

RETREATING
-3.6 to -1.0
-.9 to 1.0
1.1 to 4.0
4.1 to 6.0
6.1 to 9.0
EXPANDING 9.0 to 13.0

retreating

expanding   no change

Sonoma
Creek

Petaluma
River

Novato

SAN
PABLO
BAY

HAMILTON
FIELD

Gallinas Creek

POINT
PINOLE

San Rafael

N

POINT
SAN PABLO

1:100,000

1 mile

Richmond

Napa River

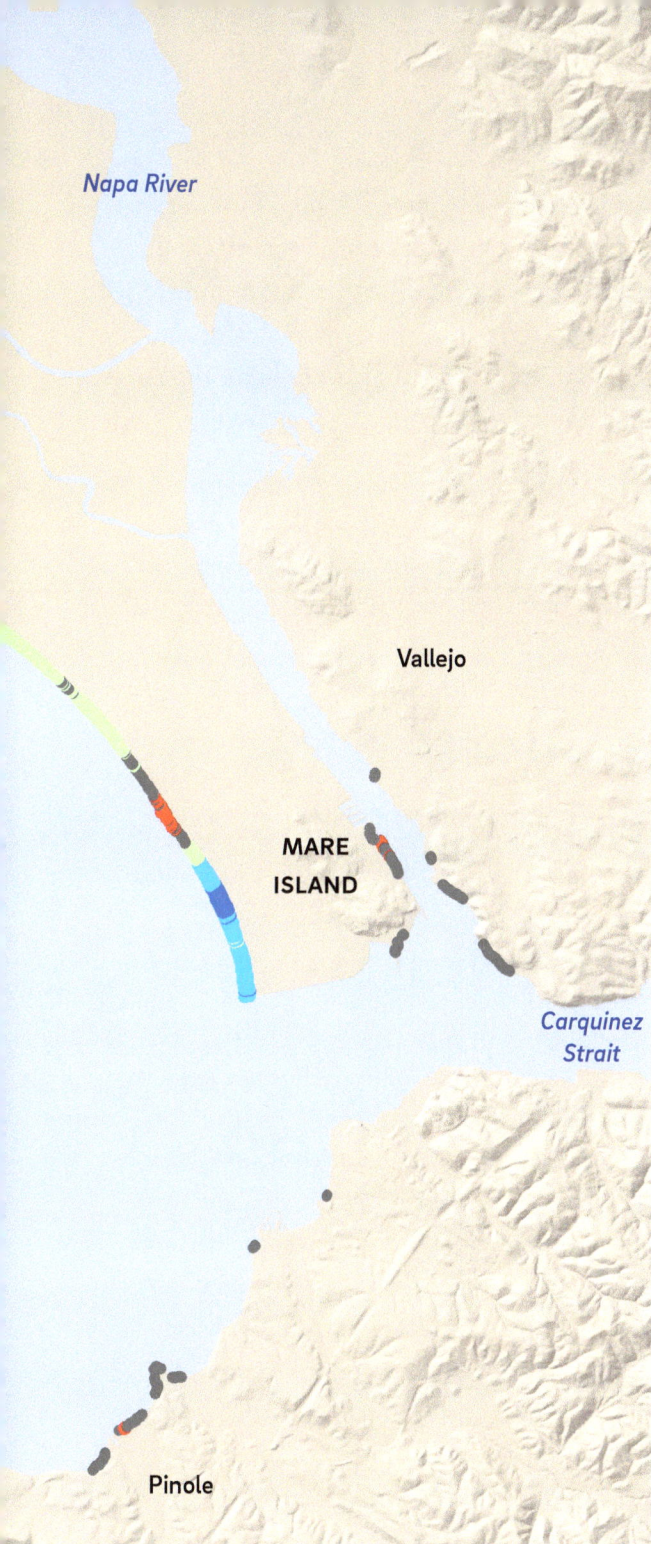

SHORT-TERM
RATES OF SHORELINE CHANGE
1993-2010

Vallejo

MARE ISLAND

Carquinez Strait

Pinole

While the long-term trend has been expansion of the shoreline around San Pablo Bay, patterns of short-term marsh evolution (1993 to 2010) are slightly more complex. More than 35% of the marsh shoreline has expanded over the last 20 years (shown in cool colors in the map), while 6% has retreated (shown in warm colors). For the majority of the shoreline length (59%), the degree of lateral change falls within the error bars of the analysis, and is considered to be statistically unchanged (shown in grey; areas calculated to have a net rate of change of between -1 and +1 m/yr are considered to be unchanged).

The most rapid marsh expansion during this time period occurred on the southern tip of Mare Island, with rates of up to 9 m/yr. There has also been a significant amount of progradation around the mouths of the Petaluma River (1-2 m/y) and Sonoma Creek (1-5 m/yr). The Marin County shoreline remained fairly static in the short-term, though the shoreline along the edge of the Hamilton Marsh restoration project juts into the Bay slightly and appears to be eroding at around 2 m/yr. The scalloped stretch of shoreline between Pt. San Pablo and the Carquinez Strait, with its unique orientation and proximity to the deep Bay outlet of the Delta, also experienced some erosion over this time period.

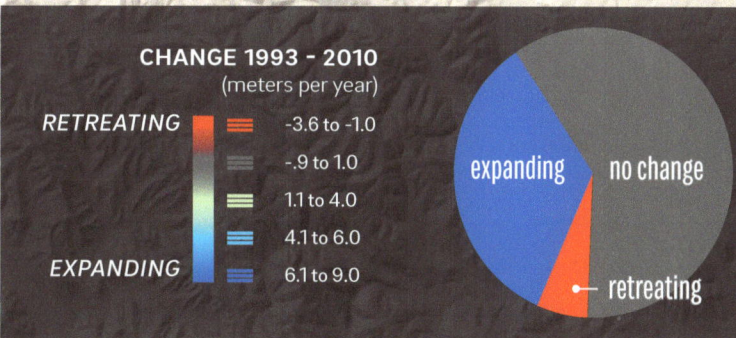

CHANGE 1993 - 2010
(meters per year)

RETREATING   -3.6 to -1.0

   -.9 to 1.0

   1.1 to 4.0

   4.1 to 6.0

EXPANDING   6.1 to 9.0

expanding   no change

retreating

33

Interpreting the short-term change

As a pilot study, this analysis is useful as a first look at the rate and direction of shoreline movement over time. In this section we examine relationships between the observed shoreline changes and the setting and morphology of the San Pablo Bay shoreline. Areas of similar setting and morphology may be subject to the same drivers, and thus may be experiencing similar patterns of change. An understanding of these relationships could help managers interpret observed changes and anticipate what type of shoreline movement might be expected in different settings.

In the absence of long-term monitoring at the site scale to observe the mechanisms by which the shoreline changes, we interpret the direction and rates of short-term shoreline movement in terms of both geomorphic settings within the Bay as well as the types of marsh edges observed along the shoreline. Although we calculated net change between three points in time, the processes which drive the erosion and progradation of the shoreline (e.g., storms, sediment availability, and many others) tend to be temporally variable, and thus the net rate of change likely obscures substantial variability in the rate of shoreline migration. Sea level rise will also change the magnitude of physical drivers affecting shoreline change.

Geomorphic units in San Pablo Bay

San Pablo Bay has repeating geomorphic patterns around the shoreline, allowing for limited observations of and comparisons between shoreline dynamics in different settings. For the purpose of this analysis, we define four major types of geomorphic settings at the shoreline (focusing on areas where marshes are found): 1) creek mouths, 2) longshore areas, 3) pocket marshes, and 4) headlands (see top right).

Several creeks and rivers (Las Gallinas, Novato, Petaluma, Sonoma, Napa, Wildcat, and San Pablo) and many smaller tributaries enter San Pablo Bay, delivering sediment to marshes, mudflats, and deep water channels. Deltas form where the creeks and rivers meet the Bay, producing distinct patterns in shoreline orientation and marsh and mudflat configurations. Between these creek mouths/deltas, there are long, exposed lengths of shoreline. Though they are oriented differently (ie., east-west, north-south), and may be convex or concave, these exposed reaches of shoreline are distinct in pattern and process from the creek mouths. Headlands jutting into the Bay represent a third context and setting for understanding shoreline movement. Point Pinole in Contra Costa County, and the stretch of shoreline between Pt. San Pedro and Gallinas Creek, provided examples of this setting. Finally, pocket marshes exist as isolated, scalloped areas of marshes between headland promontories observed mainly on the Contra Costa County shoreline.

We examined patterns in the direction of shoreline change by geomorphic unit in the Bay (right). Creek mouths were observed to have either prograded (38%) or remained stable (within the bounds of the error analysis; ~60%) between 1993 and 2010. Less than 1% of creek mouth deltas eroded over this time period. Because of their high sediment supply, concave shape, and wide fringing mudflats, creek mouths tend to represent a depositional environment. The shoreline position along headlands and pocket marshes was found to be mostly (> 85%) unchanged over this time period. Surprisingly, almost 50% of the longshore areas were found to have prograded, even though these areas of the shoreline are often characterized by a high energy, dispersive environment. About 15% of the longshore areas were found to have retreated, and the rest were unchanged.

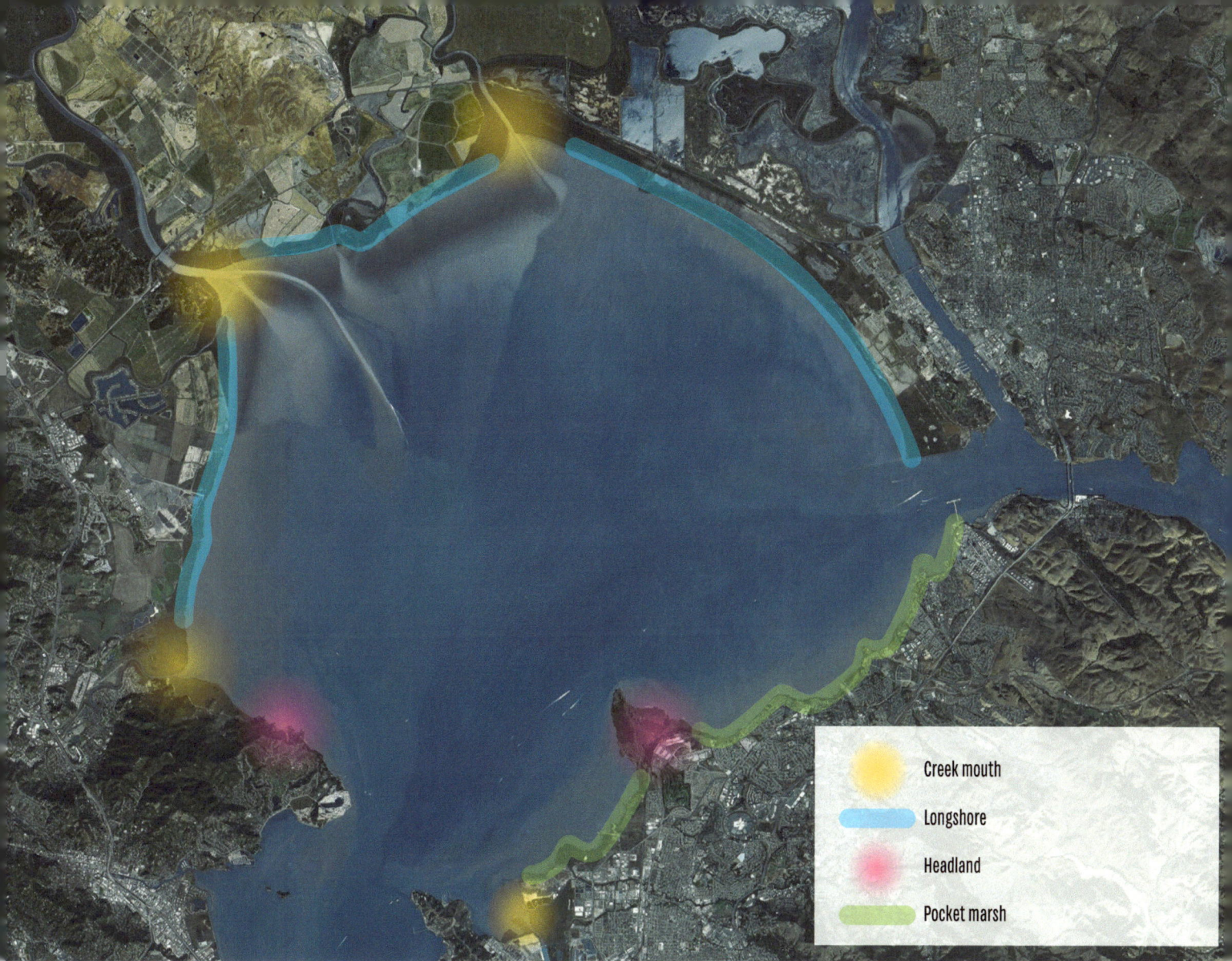

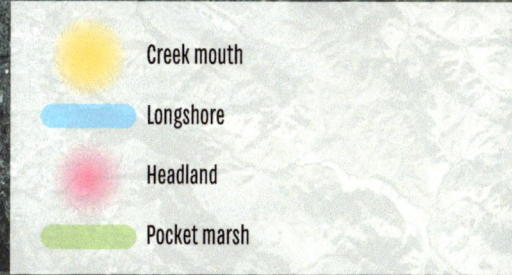

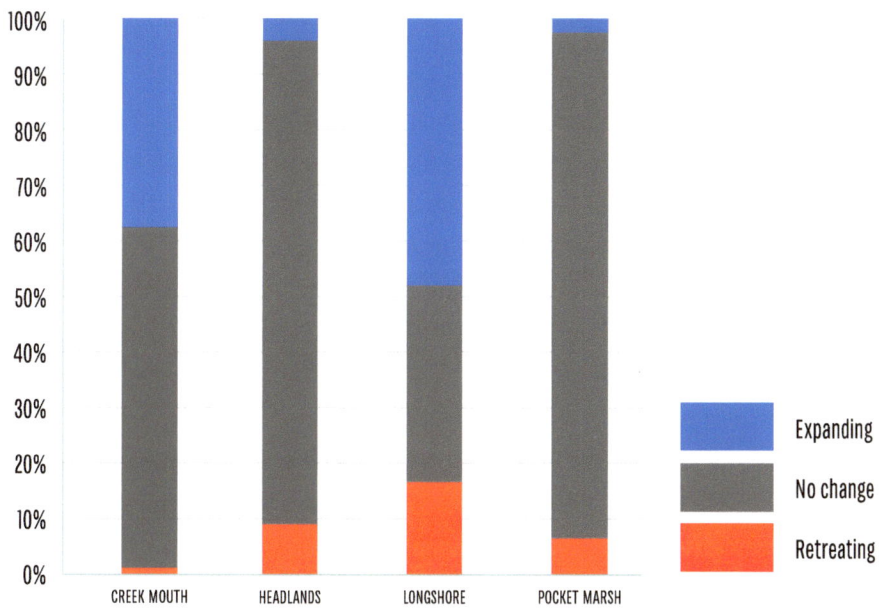

San Pablo Bay geomorphic units as defined for this study (top). To look for patterns in shoreline movement, we split the San Pablo Bay into four geomorphic units: creek mouths, longshore areas, headlands, and pocket marshes. These were defined by their proximity to creeks, topography, and shoreline orientation. (imagery courtesy ESRI)

San Pablo Bay shoreline geomorphic units by direction of change (1993-2010) (left). Creek mouths and longshore areas of San Pablo bay tended to prograde or remained stable over the short-term (grey), while marshes fronting headlands or in scalloped protected pockets were largely unchanged.

Legend:
- Creek mouth
- Longshore
- Headland
- Pocket marsh

Chart legend:
- Expanding
- No change
- Retreating

Chart x-axis: CREEK MOUTH, HEADLANDS, LONGSHORE, POCKET MARSH

Marsh edge types in San Pablo Bay

The morphology of the marsh edge can provide clues about the processes that are contributing to expansion and retreat in any given location along the shoreline. As discussed earlier (see pages 16-17), we identified five different marsh edge types as part of our mapping of San Pablo Bay's marsh shoreline (top right). In this section we analyze and interpret documented patterns of shoreline change in terms of marsh edge type, and examine the relationships between shoreline change, marsh edge type, and geomorphic setting.

As seen in the map on the right, marsh edge types vary around San Pablo Bay. We observed a fairly even distribution of these types: almost 40% of the marsh shoreline was composed of a scarp (yellow and light green), 45% was composed of ramps (brown and dark green), and almost 15% was fronted by beaches (pink).

The edge types vary geographically (discussed below) and by direction of change. While the location of almost 60% of the shoreline was statistically unchanged over the short-term, patterns still emerged when analyzing the type of marsh edge and its measured movement. Beachfronts were over 90% unchanged, most likely held in place by the headland formations which tended to back these locations often with less erodible underlying material. The ramped edges were where most of the progradation occurred. Over 70% of the ramps with no inflection points (RNI) were expanding. Over 30% of the scarps with no vegetation (SN) and 60% of the scarps with vegetation (SV) had expanded, which is counter-intuitive given their distinct vertical faces. Over 12% of the scarps without vegetation (SN) were eroding, and the majority of this marsh type was shown to be unchanged over the short-term.

However, an examination of the type of marsh edge by geomorphic unit and direction of movement over the short-term (see opposite page, and Table 2) exposes a more complex story.

About 80% of the creek mouth shorelines were characterized as ramps, and were expanding at varying rates. This may be because the marsh edges near creek mouths in San Pablo Bay are protected, as they tend to be fronted by large mudflats, and are oriented away from direct wave run-up. This can translate to lower wave heights reaching the marsh edge, supporting a ramped edge profile (Moller and Spencer 2002). These areas also may have increased sediment availability from local watersheds that can be re-suspended and increase rates of progradation. The small areas of eroding marsh edges near creek mouths were found up the creek channels. These areas were characterized as scarps without bayward vegetation (SN), which may be due to concentrated flow scouring the channel banks.

Scarps, with and without bayward vegetation, were found predominantly in longshore areas (see chart below). While scarps are often found in

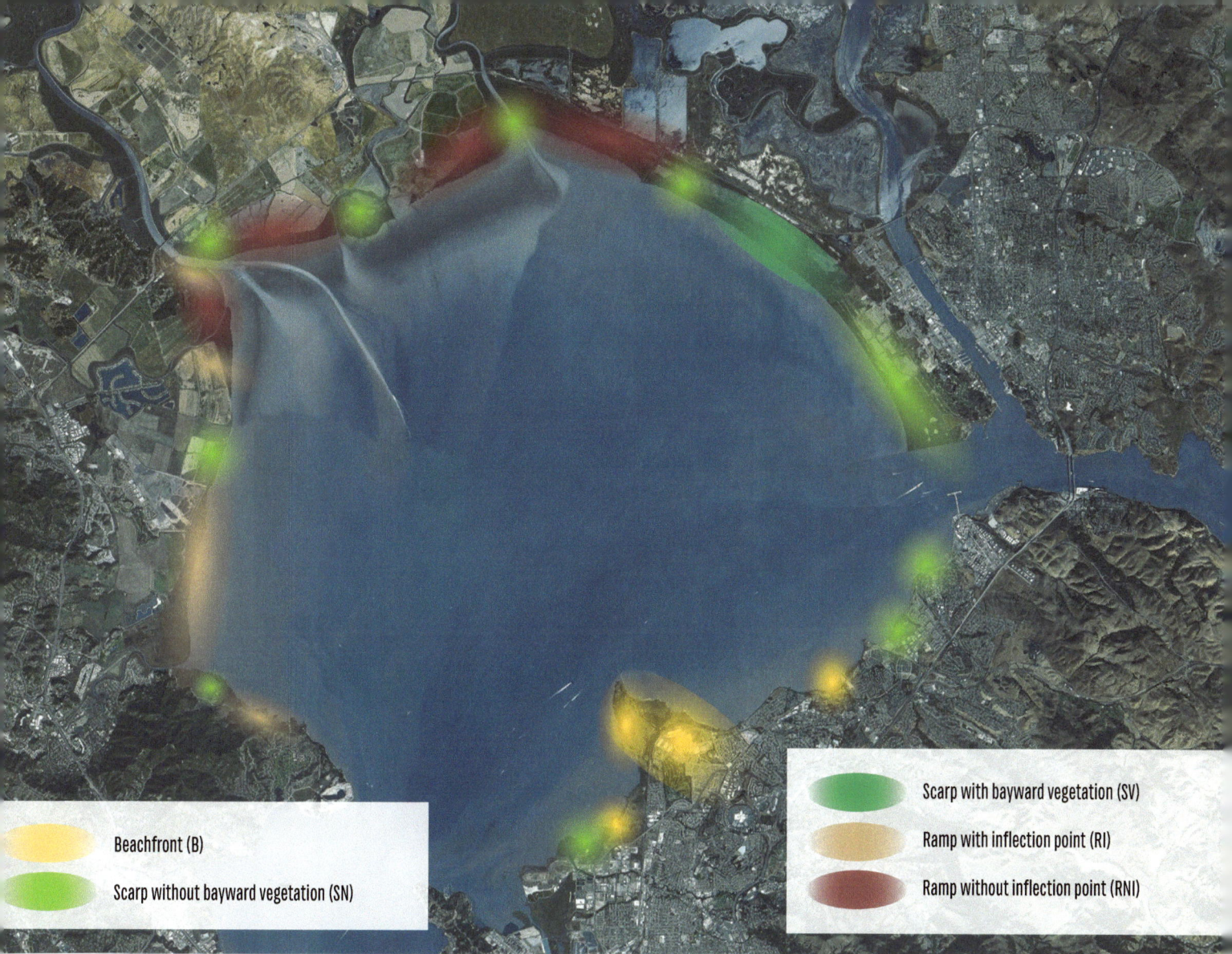

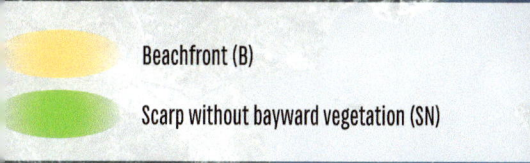

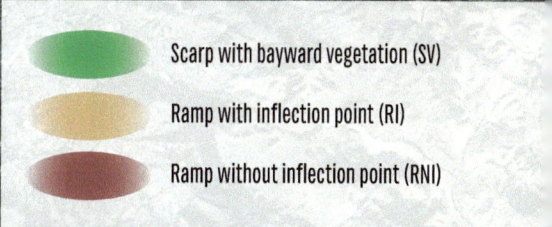

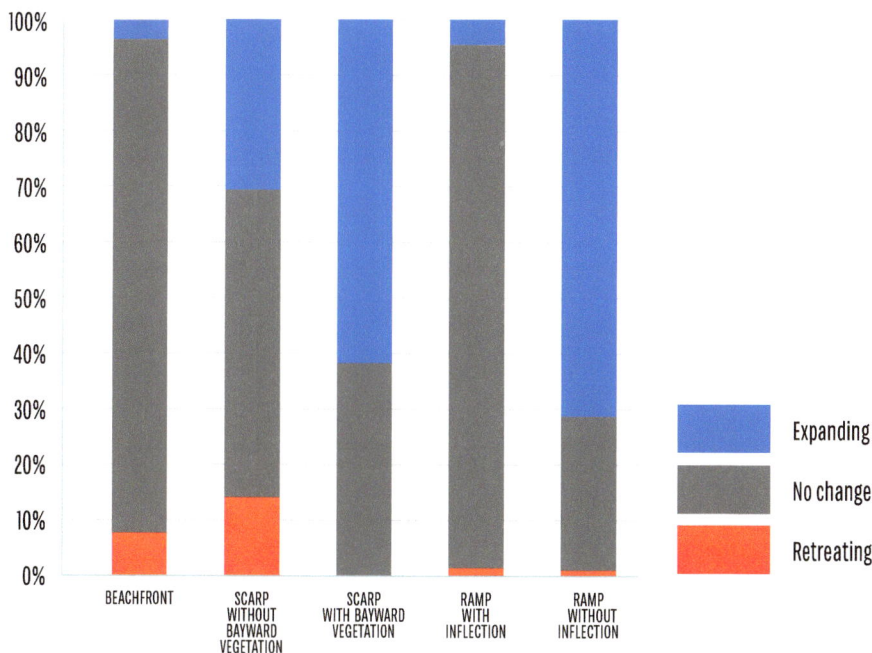

Map of San Pablo Bay showing locations of the different marsh edge types (top).

Legend:
- Beachfront (B)
- Scarp without bayward vegetation (SN)
- Scarp with bayward vegetation (SV)
- Ramp with inflection point (RI)
- Ramp without inflection point (RNI)

Map of San Pablo Bay showing locations of the different marsh edge types (top). Five different marsh edge types were distinguished while mapping the contemporary shoreline: two types of scarps (with vegetation and without), and two types of ramps (with inflection point, and without), and marshes that are fronted by beaches. These types were useful in interpreting the erosion and progradation patterns. The edge of the marsh is dynamic these types are most likely not static features. (courtesy ESRI imagery)

Marsh edge types by direction of shoreline movement (left). Over 70% of the ramps without inflection points (RNI) were shown to have expanded over the short-term, while the majority of the marsh erosion was found at the beach-fronted marshes (B), and the scarps without vegetation (SN). Surprisingly, over 60% of the scarps with vegetation in front (SV) had prograded, and the rest were unchanged. This supports the hypothesis that erosional-looking features can become sediment trapping mechanisms for prograding marshes.

high fetch, high wind wave energy and erosive environments (Moller and Spencer 2002), marsh scarps in San Pablo Bay were found to be retreating, expanding, and fairly static. Scarps with bayward vegetation (SV) were not found to be retreating at all. We found that marsh scarps had in fact prograded significantly along the "longshore" Highway 37 marsh, which may be counterintuitive, given the high wave energy environment, long fetch, and vertical scarp face of the marsh edge.

The scalloped pocket marshes were comprised of equal distributions of ramps and scarps, and also were found to be equally eroding, prograding and unchanging. Many of these areas of marsh are found between Carquinez Strait and Point Pinole, with convex shoreline positions, and varied orientations to wind wave energy. Over 95% of headlands were fronted by beaches, and over 90% of the length of beach-fronted shoreline was found to be unchanging, which may be due to the underlying topography setting often setting the shoreline position.

Geomorphic units and edge morphology seem not to be tied to the direction of shoreline movement. Other drivers such as direction with respect to wind, local sediment supply, underlying geology, and wave energy seem to be more important for determining edge conditions.

Geomorphic units of San Pablo Bay by marsh edge types (right). Over 70% of creek mouths were composed of the ramped morphology. 100% of headlands were fronted by beaches. Pocket marshes were an even mix of the marsh edge types, though over 50% were ramped. Over 70% of longshore areas were composed of scarps with no vegetation.

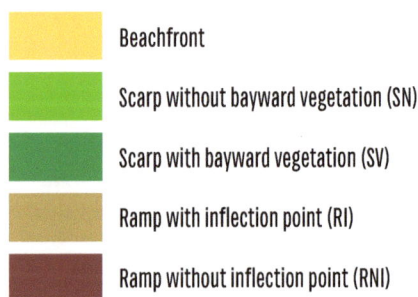

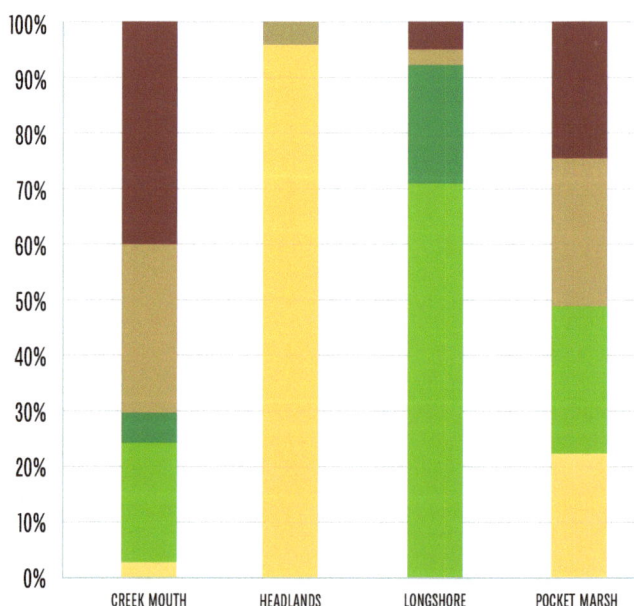

Beachfront

Scarp without bayward vegetation (SN)

Scarp with bayward vegetation (SV)

Ramp with inflection point (RI)

Ramp without inflection point (RNI)

Table 2. This table shows the total length (m) of the marsh shoreline in San Pablo Bay (1993-2010) that eroded, prograded or remained stable between 1993 an 2010, broken out by their geomorphic unit and marsh edge morphology.

(Length units in meters)		Eroding	No Change	Prograding	Total	% of total shoreline
Creek Mouth		**1400**	**105500**	**64200**	**171100**	**57%**
	B	0	4900	0	4900	2%
Scarps	SN	900	29200	6600	36700	12%
	SV	0	8100	1100	9200	3%
Ramps	RI	300	49000	2500	51800	17%
	RNI	200	14300	54000	68500	23%
Headlands		**2900**	**28700**	**1400**	**33000**	**11%**
	B	2900	28700	1400	33000	11%
Scarps	SN	0	0	0	0	0%
	SV	0	0	0	0	0%
Ramps	RI	0	0	0	0	0%
	RNI	0	0	0	0	0%
Longshore		**13100**	**28100**	**38200**	**79400**	**26%**
	B	0	0	0	0	0%
Scarps	SN	12900	20300	23200	56400	19%
	SV	0	1800	15000	16800	6%
Ramps	RI	100	2200	0	2300	1%
	RNI	100	3800	0	3900	1%
Pocket Marsh		**1100**	**15900**	**400**	**17400**	**6%**
	B	300	3600	0	3900	1%
Scarps	SN	0	4600	0	4600	2%
	SV	0	0	0	0	0%
Ramps	RI	400	4200	0	4600	2%
	RNI	400	3500	400	4300	1%
Grand Total		**18500**	**178200**	**104200**	**300900**	**100%**
% of total shoreline		**6%**	**59%**	**35%**		**100%**

Evolution hypothesis

The different marsh edge morphologies have most likely developed in response to a variety of drivers, including sediment supply and caliber, wave attack, mudflat shape and size, plant colonization patterns, and orientation to the Bay (Allen 1989, Schwimmer 2001, Moller and Spencer 2002, Pedersen and Bartholdy 2007, and others). There is likely a feedback between hydrodynamic conditions, shoreline orientation, and marsh edge typology (Allen 1989, Moller and Spencer 2002), and these typologies may represent different points along a cycle of marsh edge evolution. A hypothetical cycle of marsh edge evolution is explained on the facing page, but more monitoring must be done to test the validity of this conceptual model.

Conceptual model of marsh edge evolution (right). This conceptual model posits a hypothesis linking the four major shoreline edge types (excluding beachfronts) found in this study into an evolution story. (adapted from Allen 1989)

(Below) A photo of an eroding block from a marsh scarp. (photo by Shira Bezalel, March 2015)

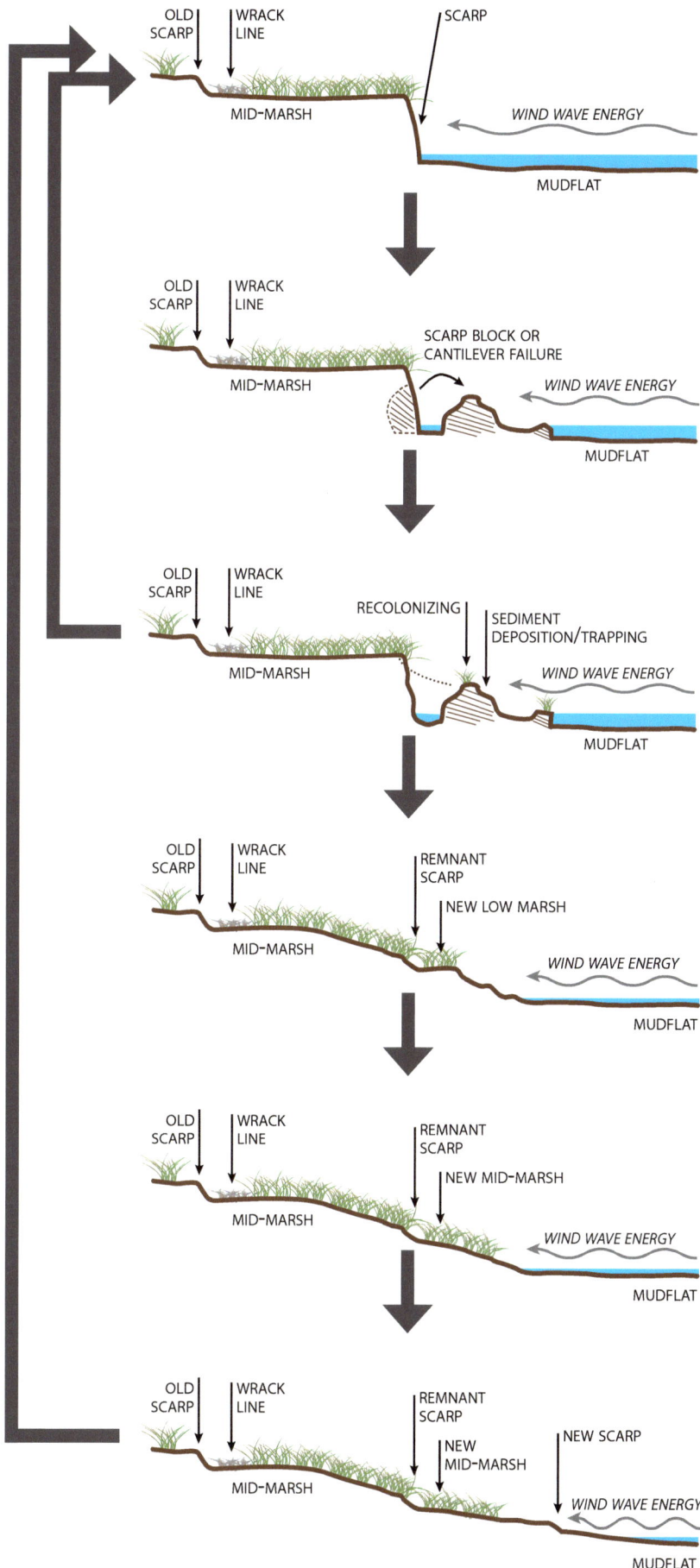

Scarp without bayward vegetation (SN)

Fails under pressure from wind wave energy or wave run-up, and undercut blocks fail or cantilever, depositing sediment (with or without vegetation) in front of the scarp.

Scarp without bayward vegetation (SN)

The failed block dissipates wave energy until this deposit is scoured away and redistributed on the mudflat or marsh plain, thus creating an erosional environment as the wave energy is then directed back to the scarp.

Scarp with bayward vegetation (SV)

If the failure is large enough to redirect wave energy for longer periods of time, the failed blocks may create an environment for sediment deposition and trapping between the old scarp and the failed block.

Ramp with inflection point (RI)

A ramped profile begins to form as sediment fills in behind the failed block, building elevation, creating new low marsh and leaving behind a remnant scarp.

Ramp without inflection point (RNI)

As the ramping continues, wave energy is dissipated such that the low marsh vegetation traps sediment, building up to mid-marsh habitat.

Ramp with new bluff forming (RI)

When the new mid-marsh levels, the ramped profile steepens and wind wave energy begins to erode the new mid-marsh, creating a new scarp. And the cycle continues...

41

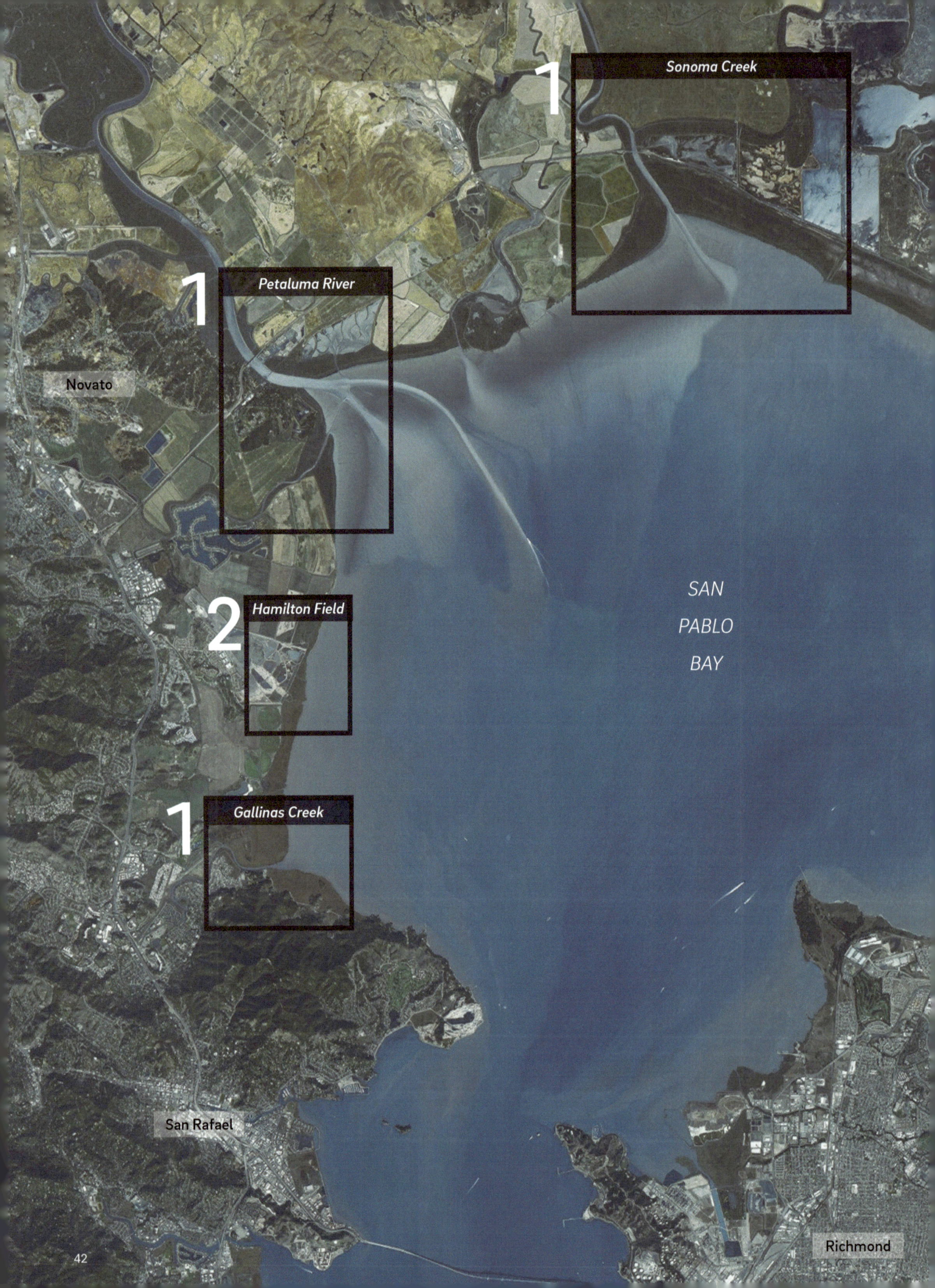

1 Sonoma Creek

1 Petaluma River

Novato

2 Hamilton Field

1 Gallinas Creek

SAN
PABLO
BAY

San Rafael

Richmond

5. PATTERNS & EXAMPLES

Several key themes emerged from our findings:

1) Much of the marsh edge is expanding baywards, and has been for 150 years, particularly around the mouths of large creeks.

2) Retreat of the marsh edge is occuring at limited locations, mainly in areas that stick out into the Bay.

3) Parts of the shoreline may look like they are eroding, but are in fact expanding rapidly.

This section is organized like an atlas, and is meant to highlight the key findings from this study, focusing on the short-term changes.

While we discuss hypotheses to explain the observed patterns of marsh expansion and retreat, further analysis should be done to determine the drivers of shoreline change in different locations.

Vallejo

Mare Island

3

Pinole

1 **EXPANDING AT CREEK MOUTHS:**
GALLINAS, PETALUMA, AND SONOMA

2 **RETREATING AT PROTRUSIONS:**
HAMILTON FIELD

3 **EXPANDING MARSH SCARPS:**
MARE ISLAND

N

1:100,000

2 miles

(imagery courtesy ESRI)

EXPANDING CREEK MOUTHS DELTAS

Marshes at mouths of creeks around San Pablo Bay continue to laterally expand

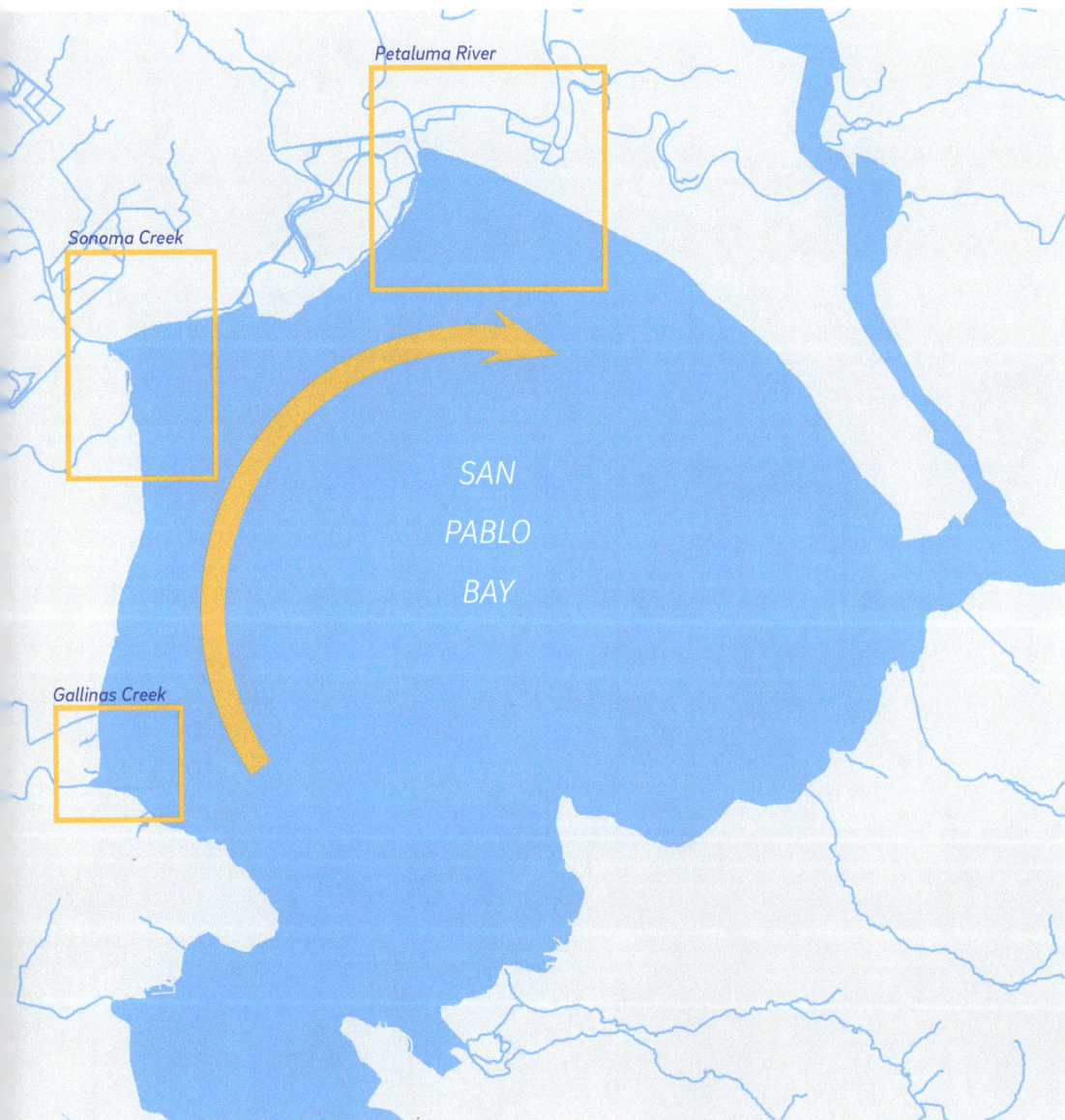

Petaluma River

Sonoma Creek

SAN
PABLO
BAY

Gallinas Creek

OVERVIEW We found that more than 38% of the shoreline around creek mouths in San Pablo Bay has experienced net expansion over the past 20 years. Short-term (1993-2010) and long-term (ca. 1855-1993) expansion rates are in many cases nearly comparable (on the order of 1-3 m/yr), though the amounts and rates of expansion vary by location. Traveling clockwise from Gallinas Creek and following the path of the San Pablo Bay Gyre, there is an increase in both local sediment supply and the rate of expansion at creek mouths (Table 3). Gallinas Creek shows no change (or change below our detection levels in the 20 year short term time period), while Novato, Petaluma, and Sonoma creeks continue to prograde. The majority (60%) of the shoreline by length in this geomorphic class is shown to be not significantly changing, though further study is needed.

PROCESS A combination of wind-wave energy and direction, local sediment supply and caliber, plant colonization and root strength, and sea level rise controls the evolution of the marsh edge (Allen 1989, Schwimmer and Pizzuto 2000, Schwimmer 2001, Moller and Spencer 2002, Pedersen and Bartholdy 2007 and others). At a local scale, progradation is often (but not always) observed in areas oriented away from direct wave runup, or in areas where high local sediment supply and local bathymetric trends have created long, wide mudflats, such as the mouths of Petaluma River and Sonoma Creek. Most likely this is due to a mix of sediment delivery from local watersheds (Table 3), and reworked sediment from nearshore mudflats (Gunnell et al. 2013). These advancing marsh edges mainly occur in areas characterized by ramps with no inflection point.

MANAGEMENT IMPLICATIONS Expanding marshes at creek deltas may offer some important opportunities for the green infrastructure, complete marsh habitat, carbon sinks, and other benefits, especially in light of sea level rise. For example, Collins et al. (1994) found Ridgway's rail (Rallus obsoletus) counts to be highest at creek mouths (Collins et al. 1994). Knowing where and at what rate marshes are expanding, and the drivers or progradation is critical when determining where to place sea level rise adaptation strategies. Enhancing local sediment supply and reconnecting creeks to their marshes may increase the likelihood that marsh accretion and progradation will keep pace with sea level rise, at least in the short term.

Table 3. **Local estimate of sediment supply from major watersheds** entering San Pablo Bay, clockwise from southwest to northeast (from McKee et al. 2013), and average rates of change of adjacent shoreline.

Watershed	Total watershed area (km²)	Average suspended sediment (1995-2010)*		Range of marsh edge rate of change 1993-2010 (m/yr)
		metric tons	metric tons/km²/yr	
Gallinas Creek	14.5	Unknown	Unknown	(-) 0.9 to (+) 1.0
Novato Creek	96.2	7366	77	(-) 1.0 to (+) 1.5
Petaluma River	122	26059	213	(+) 1.1 to (+) 2.0
Sonoma Creek	241	204516	847	(+) 2.1 to (+) 5.1
Napa River**	738	310928	422	(+) 4.1 to (+) 7.0

*Mckee et al 2013. **Napa River rates of change refers to the jetty at the edge of Mare Island

Sonoma Creek, imagery courtesy Google Earth

EXPANDING CREEK MOUTH
GALLINAS CREEK

At the mouth of Gallinas Creek, China Camp and McInnis marshes were found to have expanded minimally between 1993 and 2010, with although the migration rates were between -1.0 and 1.0 m/yr, and thus within the error margins of our study. The marsh edge here is primarily characterized as a ramped profile with an inflection point (RI), with one stretch on the southwest side of Gallinas Creek that is observed to be an unvegetated scarp (SN). This area is mapped with medium uncertainty because of the quality of the aerial imagery.

Gallinas Creek Mouth in 2010 (left). (2010 NAIP); Gallinas Creek mouth in 1974 (above) [courtesy of Marin History Museum]

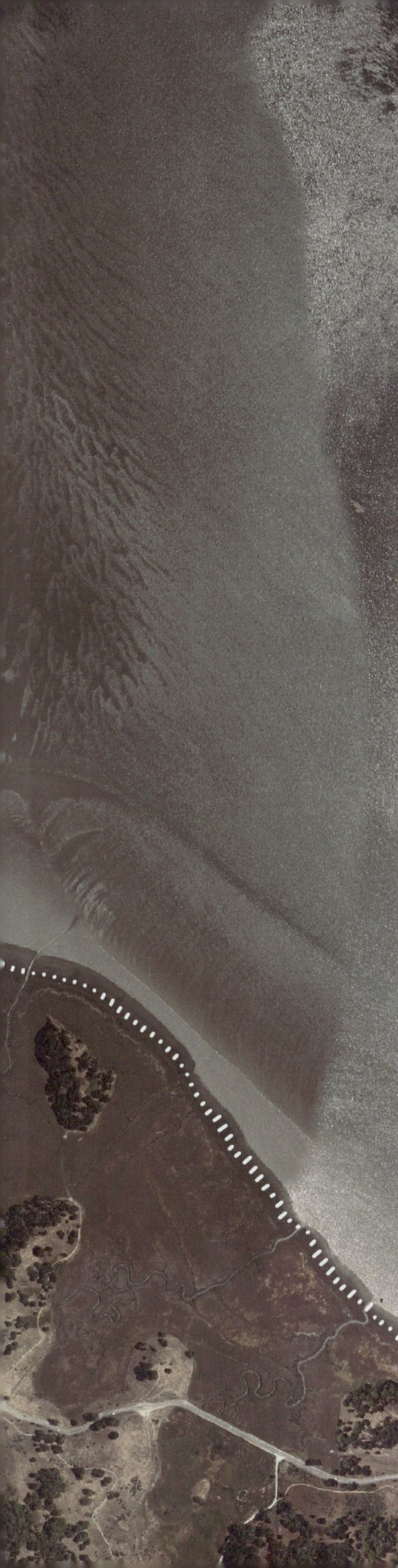

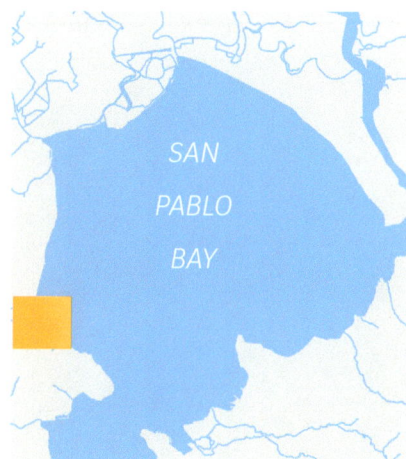

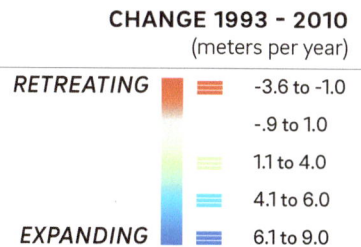

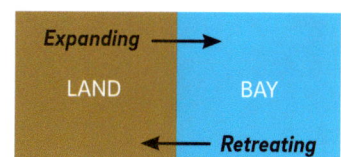

SAN PABLO BAY

CHANGE 1993 - 2010
(meters per year)

RETREATING -3.6 to -1.0
-.9 to 1.0
1.1 to 4.0
4.1 to 6.0
EXPANDING 6.1 to 9.0

Expanding →
LAND   BAY
← Retreating

1:20,000

.25 mile

N

EXPANDING CREEK MOUTH
PETALUMA RIVER

The marshes at the delta of the Petaluma River have been expanding consistently at 1-2 m/yr between ca. 1855 and 2010 (spanning both time periods of the study). The edges facing the Bay are characterized by a ramped morphology with no inflection point (RNI). Wide mudflats extend across this shallow embayment and most likely drive resuspension of sediment that then is directed at the marsh edges, contributing to the progradational trend. We noted a small area of short-term erosion on the outlet to the Sonoma Baylands estoration project (right, outlined in red). This may have to do with scour associated with increased tidal action in the relatively new channel breached through the marsh plain.

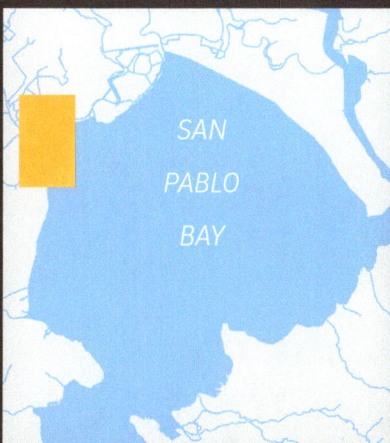

CHANGE 1993 - 2010
(meters per year)

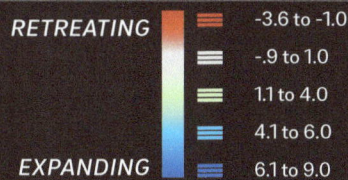

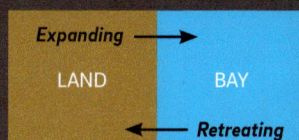

RETREATING	-3.6 to -1.0
	-.9 to 1.0
	1.1 to 4.0
	4.1 to 6.0
EXPANDING	6.1 to 9.0

Expanding →
LAND   BAY
← Retreating

Petaluma Creek mouth from above (top). These images highlight the main tidal channel as it enters San Pablo Bay, as well as the complexities of the marsh edge. The low marsh (unmapped in this study) is visible in green, in contrast to the brown marsh plain. The mouth to the Sonoma Baylands channel is outlined in red in the three aerial photos. Small amounts of erosion were noted here. ([top] courtesy Wikipedia; [bottom] photo by Micha Salomon, February 2012; [left] 2010 NAIP imagery)

N

1:20,000

.25 mile

50

CHANGE 1993 - 2010
(meters per year)

RETREATING ▬ -3.6 to -1.0
-.9 to 1.0
1.1 to 4.0
4.1 to 6.0
EXPANDING ▬ 6.1 to 9.0

Expanding →
LAND   BAY
← Retreating

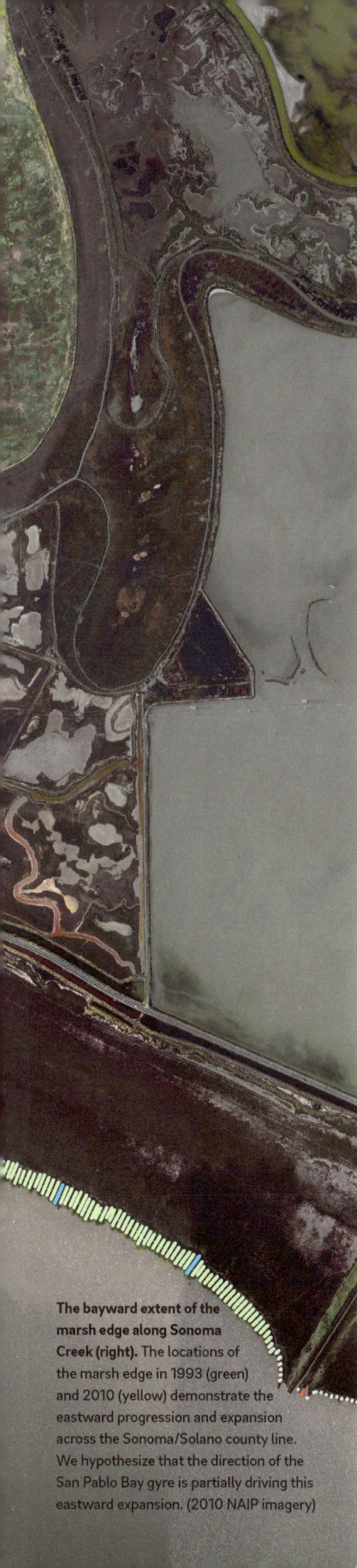

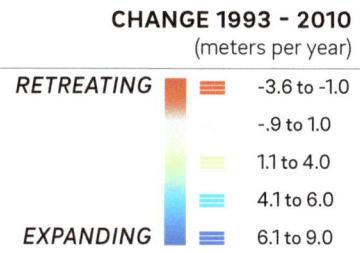

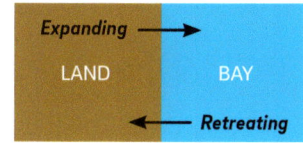

EXPANDING CREEK MOUTH
SONOMA CREEK

The marshes fronting Sonoma Creek have been expanding at a relatively fast pace over both the long- and short-term. Between ca. 1855 and 1993, both sides of the creek expanded at an average rate of 4-5 m/yr, slowing to between 1-3 m/yr between 1993 and 2010. The western lobe of the Sonoma Creek delta has been both expanding outwards and migrating eastward, slowly squeezing the width of the creek outlet (below). The outer edges of the marsh are characterized as ramps with no inflection point (RNI), which is perhaps indicative of their quickly prograding nature. Like the delta at the mouth of the Petaluma River, Sonoma Creek has developed wide mudflats extending into San Pablo Bay. The orientation of the subtidal delta and the shoreline are also more protected than the longshore areas along Mare Island, which, combined with wide shoal-inducing mudflats, may contribute to the high rates of progradation along this stretch of shoreline.

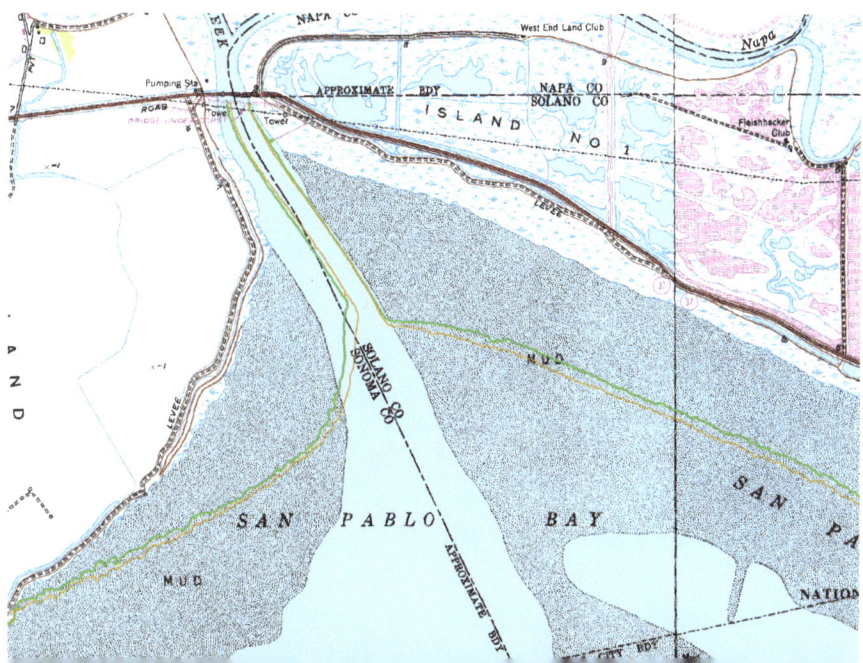

The bayward extent of the marsh edge along Sonoma Creek (right). The locations of the marsh edge in 1993 (green) and 2010 (yellow) demonstrate the eastward progression and expansion across the Sonoma/Solano county line. We hypothesize that the direction of the San Pablo Bay gyre is partially driving this eastward expansion. (2010 NAIP imagery)

RETREATING AT PROTRUSIONS

Marshes near artificial protrusions in San Pablo Bay are eroding

SAN

PABLO

BAY

Hamilton Field

OVERVIEW   Only 6% of the mapped marsh shoreline in San Pablo Bay experienced net erosion between 1993 and 2010. Erosion was concentrated in three hotspots: 1) in Marin County near the Hamilton Marsh restoration site, 2) in Sonoma County at the mouth of Tolay Creek at Midshipman's Point , and 3) in Solano County on the south side of Mare Island. All three of these areas are located in longshore settings, and are characterized by human-made protrusions; for example, the erosion in Marin occurs near the apex of the built-out and filled Novato Baylands. The rates of erosion in these three hot spots vary between -1 and -3.5 m/yr.

PROCESS   Protrusions in the longshore areas of the shoreline bare the brunt of wave attack, and therefore are subject to higher rates of erosion than other parts of the shoreline (Schwimmer 2001, Francalanci et al. 2011). In addition to direct wind wave attack, other mechanisms of erosion can include cantilever failure and filling, cracking due to wetting and drying (Allen 1989), creek mouth widening (van der Wal and Pye 2004), neck cut-offs, scalloping, and undercutting (Allen 1989). Erosion can occur both chronically and episodically (Schwimmer 2001), depending on the orientation of the shoreline, the stratigraphy of the scarp, and the presence of locally protruding or scalloping marsh surfaces (which effectively shield adjacent mudflats from erosive forces; Fagherazzi 2013, Gunnell et al. 2013). Further studies are needed to identify the relative importance of the various mechanisms and the timing of erosion in different areas of the shoreline.

MANAGEMENT IMPLICATIONS   The common perception is that marshes are singularly and uniformly eroding around the Bay, but our findings show that erosion in San Pablo Bay is concentrated at localized hotspots. Identification of these erosional hotspots can help managers to prioritize where to implement adaptation strategies such as mudflat recharge, increased sediment supply, or realignment of the shoreline orientation (where it is constrained by a levee). Eroding hotspots may also indicate areas to focus on land acquisition for transition zone or upland buffer areas. However, in certain places, it may be determined that a retreating shoreline should be abandoned and resources allocated elsewhere.

In some locations, partial or full beach creation or augmentation could be used to protect retreating marsh edges. For example, beaches historically fronted marshes along Point Pinole (and in some cases still do), and these could be restored to protect the currently eroding marsh (below). Understanding the dynamics and evolution of the marsh edge over the long and short term can help with matching appropriate management strategies to appropriate locations.

——— ca. 1855 shoreline

——— 1993 shoreline

——— 2010 shoreline

Eroding marshes (left). Though beaches still front many of the marshes on Point Pinole, there are a few locations which historically supported beaches but have continually been receding. Today, this area is an eroding marsh scarp. It might be beneficial to consider beach restoration projects in similar locations. **(top left)** Example of a neck cut-off erosional feature along the same area in Point Pinole. (photo by Shira Bezalel, February 2015). **(below)** Eroding Hamilton Field. (imagery courtesy Google Earth)

In 2014, this stretch of the marsh was breached to re-connect the former wetlands at Hamilton Army Airfield to the Bay. The breach is almost 300 feet wide, and has opened 648 acres up to tidal action at high tide. (photo by Shira Bezalel April 2015)

N

1:10,000

.25 mile

2010 NAIP imagery

RETREATING AT PROTRUSIONS
HAMILTON FIELD

CHANGE 1993 - 2010
(meters per year)

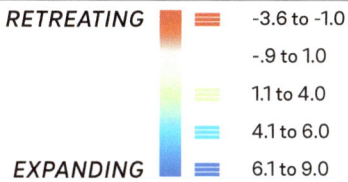

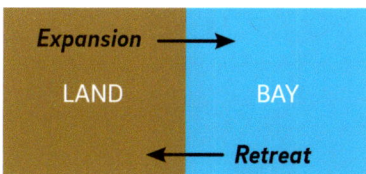

RETREATING		-3.6 to -1.0
		-.9 to 1.0
		1.1 to 4.0
		4.1 to 6.0
EXPANDING		6.1 to 9.0

Double bench profile and oblique imagery along the Novato shore (right). As most of the marsh scarps observed in San Pablo Bay had one bench, it was unusual to find a double bench in both the LiDAR and the aerial imagery. This could be a result of a particularly dynamic shoreline, receding and expanding at various time scales. More monitoring should be done to understands the mechanisms at work. (photos by Shira Bezalel [above] April 2015, and Micha Salomon [right] February 2012)

One of the few erosional locations in San Pablo Bay centers on the stretch of shoreline near the Hamilton wetlands restoration site. Between ca. 1855 and 1993, this area advanced extremely rapidly (>8 m/yr), with marshes expanding out and building up on the mudflats. In recent decades (1993-2010), however, the shoreline has been retreating at rates of -2 to -3 m/yr, in part due to its position closer to deeper water and the absence of wide mudflats. The marsh edge is characterized as a scarp without vegetation fronting it (SN), but it was also observed to be have a "double bench" profile (below).

In 2014, this area was breached as part of the Hamilton wetlands restoration project. Further monitoring along this part of the shoreline should be continued, especially given the investment put into the restoration project.

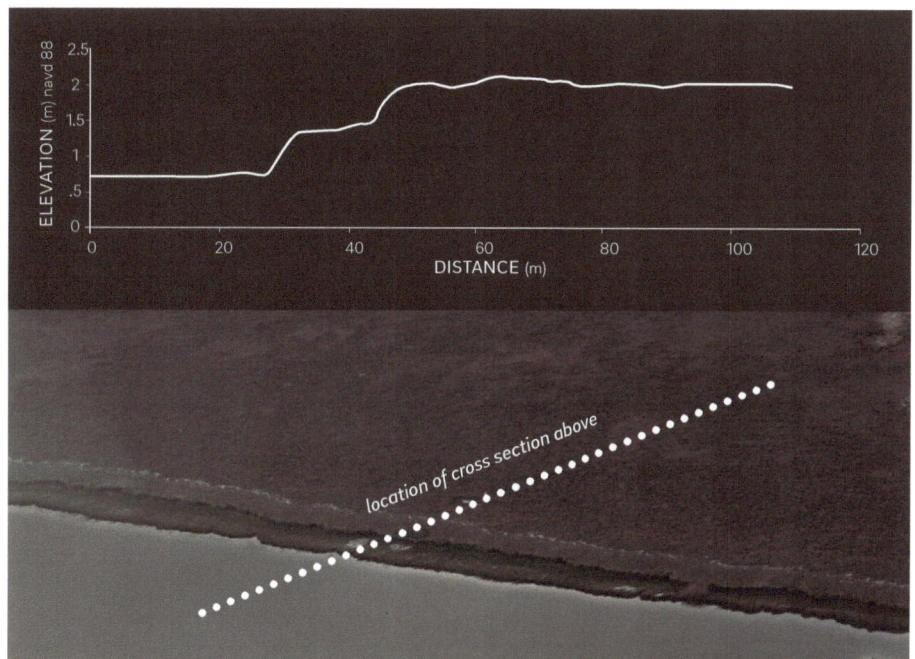

EXPANDING SCARPS

landscapes features that appear erosional but are expanding

SAN

PABLO

BAY

Mare Island

OVERVIEW   The Mare Island marsh complex (also known as the Highway 37 marsh) stretches from the tip of Mare Island northwest to the mouth of Sonoma Creek (separated by a borrow ditch and levee system). This part of the shoreline has expanded as much as 1 km in some locations since 1855. The marsh edge was historically a wave-built overwash terrace, which drained away from the Bay into marsh islands between the Napa River and Sonoma Creek (Atwater et al. 1979). The wave-built berm provided relatively high ground upon which Highway 37 was constructed in the early 20th century. Today, the edge of the marsh is still a wave-built berm, but the marsh is disconnected by Highway 37 from the baylands behind it.

This stretch of marsh provides an interesting case study to examine surprising marsh edge dynamics. The shoreline is characterized by 1-2 m high scarps, some with vegetation in front (SV) and some without (SN). A casual observer might assume that this area is eroding, because other areas with a similar scarp morphology in the Bay are eroding. However, this part of the marsh is prograding at the fastest rates observed in the study area (up to 8 m/yr in some places).

PROCESS This setting experiences high wave energy as a result of a long fetch and exposed shoreline, which maintains the scarp formation at the marsh edge. Areas with high wave energy and scarp edge types are often associated with shoreline erosion, but several factors combine to make this an area of long-term expansion. Wide mudflats provide a nearby sediment sink and help buffer wave attack. In addition, a jetty at the southern end of Mare Island stretches 1.6 km towards the center of San Pablo Bay (US Coast and Geodetic Survey 1917). This jetty functions like a beach groin at the end of the San Pablo Bay Gyre, trapping and settling the coarser sediment. The high wave energy resuspends the sediment trapped by the jetty, which then may be redeposited in the shallow nearshore environment. Thus, the high wave energy maintains the scarp at the shoreline, while providing a source of suspended sediment that allows for expansion of the marsh. We present a conceptual model which might explain this process on page 41.

MANAGEMENT IMPLICATIONS

This particular setting drives the wave-built berm morphology and the rapid expansion of marsh into San Pablo Bay, which may help explain the lack of classic sinuous tidal channels in this area. This area has been considered a poor-functioning marshland, and a management problem, but this high marsh area is rare, and should not be made to look like other marshes which are driven by difference processes.

Marsh edge morphologies often indicate direction of change. Ramped marsh faces usually are associated with prograding marshes. However, we find that marsh scarps do not necessarily indicate erosion. In some areas, such as the Mare Island example and in other wave-built berm environments (possibly in Suisun Bay, and near Novato Creek), the combination of high wave energy and a high marsh draining away from the Bay may support areas that look erosional at the marsh edge but are in fact prograding.

Mare Island. ([top left] photo by Shira Bezalel, June 2015; [below] courtesy Google Earth)

EXPANDING SCARPS
SOUTH MARE ISLAND

These images show an erosional looking feature, which is expanding, or building outwards. In the image on the left, the marsh scarp is bare, with some vegetation establishing on its bayward side. In the image on the right, the area between the scarp and the established vegetation is beginning to fill in. See page 41 for full explanation of conceptual model of expanding erosional features. (photos by Micha Salomon, February 2012)

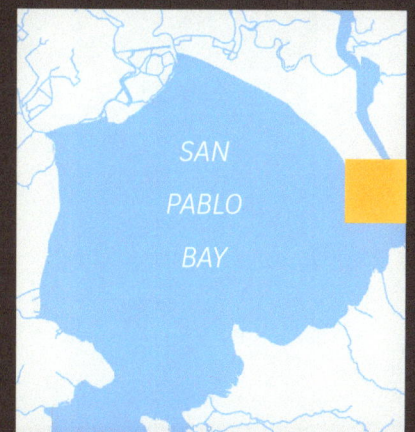

SAN PABLO BAY

CHANGE 1993 - 2010
(meters per year)

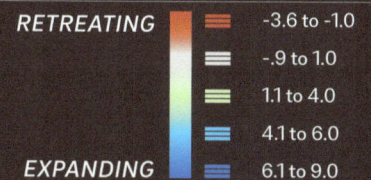

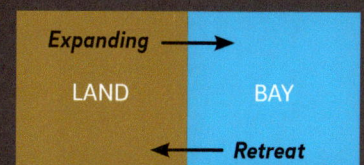

RETREATING -3.6 to -1.0
              -.9 to 1.0
              1.1 to 4.0
              4.1 to 6.0
EXPANDING   6.1 to 9.0

Expanding →
LAND   BAY
← *Retreat*

In the 1851 T-sheet **(above)** Mare Island was still an actual island. The high ground met the Bay at the southern end where the jetty, constructed in 1917, may have added to the rapid outward expansion of marsh along this shoreline. The southern tip of the island is shown in a dashed brown line on both the T-sheet and 2010 imagery (Rogers 1856, NAIP 2010).

N

1:20,000

.25 mile

2010 NAIP imagery

6. CONCLUSION

Tidal marshes and mudflats within the intertidal zone are particularly vulnerable to erosional processes. A combination of natural and anthropogenic processes is leading to a widespread loss of these critical ecosystems in many parts of the world, a trend that will likely increase with sea level rise (McLaughlin et al. 2015). Conversely, this study found that, over the last two decades, more of the marshes in San Pablo Bay have expanded (35% by length) than retreated (6%). Most of the expansion was centered around creek mouths, indicating that local watersheds, and their connection to marshes, may be critical for marsh persistence. It remains to be seen whether this positive trend will continue with accelerated rates of sea level rise, a local sediment deficit, and an eroding subtidal environment (Jaffe et al. 2007).

Marshes provide several functions, including wildlife habitat, filtering pollutants, providing flood protection, and buffers from storm surges, and many others. As opposed to levees and flood walls, which intensify wave energy, mudflats and marshes act as energy dissipators (Moller and Spencer 2002), lowering wave height and attenuating wave energy, while providing many other ecosystem services. With increased storminess and rising seas however, the marshes that remain around San Pablo Bay and within the larger San Francisco Estuary will provide increased benefits, but will also be a greater challenge to maintain.

Marshes are dynamic and can evolve in three directions: vertically, upslope, and laterally (Brinson et al. 1995). Vertical accretion rates relative to sea level rise, and elevation capital of existing marshes, are often presumed to be the primary indicators of marsh survival (Kirwan et al. 2010), but several recent studies have suggested that marsh loss rates associated with sea level rise could be more impacted by marsh edge erosion than by marsh drowning (Kirwan et al. 2010, Mariotti and Fagherazzi 2013, McLaughlin et al. 2015). In addition, inland migration of the landward marsh edge is often hindered by infrastructure, making the resilience of the bayward edge even more important. Recent studies have found that lateral expansion of the marsh can be a very fast process when sediment supply is available (Gunnell et al. 2013, Fagherazzi et al. 2013), and lateral erosion of the marsh, though slower, can be on the order of 1-3 m/yr. This is consistent with our findings. Finally, where erosive processes dominate, acquisition of upland transition zone will be necessary to allow marshes to migrate upslope with sea level rise.

In many cases, the direction of lateral expansion and retreat of the marsh edge can be impacted by choices made by the management community. Marsh expansion and retreat directly affects marsh width, which determines potential for wave attenuation and other ecosystem services we depend on. It is becoming increasingly important to understand this third direction of marsh evolution as a key indicator of shoreline resilience, as data inputs for sea level rise models (Mudd 2011), and as a method for assessing an important piece of the puzzle of prioritizing restoration adaptation strategies.

Next steps

As a pilot project, mapping shoreline change in San Pablo Bay was useful to begin to understand shoreline dynamics in the San Francisco Bay.

- It will be important to understand the variation and controls on marsh edge dynamics at a regional scale. Mapping of the marsh edge and further developing change over time metrics should be completed for the entire Estuary.

- Furthermore, as change was shown to be occurring on an order of 1-2 m/yr, it will be important to develop a systematic monitoring strategy, with remote and field components, to track trends at the appropriate frequency and resolution.

- Part of this understanding will come from confirming the conceptual model of the marsh scarp evolution presented here though field observations and modeling. This will help prioritize appropriate adaptation measures, such as coarse beach restoration, in reasonable locations.

- To fully understand the potential for continued change, and to guide management, it will be necessary to quantify conceptual models of marsh evolution and response to changes in sea levels and sediment availability. This could include pairing rates of lateral change with wind wave energy, mudflat width and slope, sediment availability data and local rates of accretion or subsidence, to get a full picture of drivers controlling marsh survival.

Photo by Julie Beagle, June 2015

Appendix A: Mapping Uncertainty

Georeferencing uncertainty (U_g) "refer(s) to the elected maximum acceptable error" for georeferenced maps (Hapke et al. 2011). This value was not available for the T-sheets that were originally used to map the Historical Baylands layer, but we estimate this to be 50 m for the ca. 1855 shoreline. U_g is not applicable to the 1993 and 2010 shorelines because georeferenced maps were not used to digitize the shoreline.

Air photo uncertainty (U_a) applies to the uncertainty contained in the mapping sources of the 1993 and 2010 shorelines. The metadata lists these values as ±7 m for the 1993 DOQQ imagery and ±6 m for the 2010 NAIP imagery. In practice this value is likely lower, as such imagery is generally more accurate in the typically flat areas near the shoreline than in hilly or mountainous areas that are outside of the study area. Because of this fact, and our observations while digitizing, we used the same value as Hapke et al. (2011): ±3 m.

T-sheet uncertainty (U_t) refers to the uncertainty from the original T-sheet mapping process in the 1850s. This value is only applicable to the ca. 1855 shoreline and is set at ±10m (Shalowitz 1964 in Hapke et al. 2011).

Digitizing uncertainty (U_d) applies to error introduced by the mapper during the digitizing process. For the ca. 1855 shoreline the digitizing error was set at ±1 m following Hapke et al. (2011). For the 1993 and 2010 shorelines, this was less straightforward. The mid-marsh/low-marsh boundary is not equally apparent across the reference imagery. Depending on the clarity of the mappable shoreline in the digitizing imagery, values for digitizing uncertainty were set at 2 m,

10 m or 20 m. A 1 m value was applied only to the 2010 shoreline where the shoreline was entirely clear. In two-thirds of these cases the location of the shoreline was confirmed by a high resolution 2 m Lidar-derived hillshade. In some cases there was little doubt as to the shoreline position for these segments. In other areas, the shoreline was blurred to some extent, so the shoreline position was mapped using best professional judgement. The blurring likely indicates pixels that are a mixture of open water, bay mud, and sparse vegetation. An U_d value of ±10 m was assigned to the remaining 2010 shoreline. Segments of the 1993 shoreline were assigned U_d values of ±10 m or ±20 m depending on how easily discernible, or conversely how blurred, the area immediately surrounding the mapped shoreline was (refer to examples shown in above section) (Table 4).

The overall location uncertainty value is used by DSAS to automatically calculate the EPR uncertainty for each transect using this equation (Himmelstoss 2009):

Overall Certainty$=\sqrt{[(\text{Uncy A})^2+(\text{Uncy B})^2]}/(\text{date A}-\text{date B})$

Where 'uncy A' and 'uncy B' are the positional uncertainties of the shoreline segments intersected by the transect, and dates A and B are the dates of the shorelines. Using this approach, for the ca. 1855 and 1993 rates, approximately 18% of the transects had uncertainty rates greater than the rate of movement. For 1993 and 2010 rates, approximately 57% of the transects had rate uncertainty greater than the rate of movement. Because of this uncertainty, we considered areas between ±20 m to be "no change" since the uncertainty exceeds the change in shoreline position that may have occurred.

Table 4. Uncertainty type as applied to the shorelines mapped for this project. The USGS values are noted for reference.

Uncertainty Type	Hapke et al. 2010	1855 SFEI (EcoAtlas)	1993 SFEI (DOQQ)	2010 SFEI (NAIP)	Notes
U_g Georeferencing	Tsheets only ±4 m	?	NA	NA	EcoAtlas: Legacy dataset derived from Tsheets was used. T-sheets were georeferenced pre-GIS
U_d Digitizing	past studies ±1 m	±50 m	±10 m or ±20 m	±2 m, ±10 m or ±20 m	Often unidirectional (e.g. shoreline might be further bayward but not landward)
U_t Tsheet	Shalowitz ±10 m	±10 m	NA	NA	
U_a Air photo Uncy	±3 m assigned	NA	±3	±3	1993 DOQQ accuracy = ± 7.6 m at 95% confidence interval; 2009 NAIP reported horizontal accuracy <=6 m

REFERENCES

Allen, J. R. L. 1989. Evolution of salt-marsh cliffs in muddy and sandy systems: a qualitative comparison of British west-coast estuaries. *Earth Surface Processes and Landforms* 14, 1: 85-92.

Atwater, B.F. 1979. Ancient processes at the site of southern San Francisco Bay: movement of the crust and changes in sea level, in San Francisco Bay: the urbanized estuary, T.J. Conomos (ed), Pacific Division, American Association for the Advancement of Science, San Francisco, California.

Atwater, B. F., Conard, S. G., Dowden, J. N., Hedel, C. W., & MacDonald, R. L. 1979. History, landforms, and vegetation of the estuary's tidal marshes in San Francisco Bay: the urbanized estuary, T.J. Conomos (ed), Pacific Division, American Association for the Advancement of Science, San Francisco, California.

Barnard, P. L., Schoellhamer, D. H., Jaffe, B. E., & McKee, L. J. 2013. Sediment transport in the San Francisco Bay coastal system: An overview. Marine Geology, 345, 3-17.

Bay Conservation and Development Commission (BCDC). 2013. Corte Maderal Baylands Conceptual Sea Level Rise Strategy. Prepared by BCDC and ESA-PWA.

Bever, A. J., & MacWilliams, M. L. 2013. Simulating sediment transport processes in San Pablo Bay using coupled hydrodynamic, wave, and sediment transport models. Marine Geology, 345, 235-253.

Brinson, M. M., Christian, R. R., & Blum, L. K. 1995. Multiple states in the sea-level induced transition from terrestrial forest to estuary. Estuaries, 18(4), 648-659.

Cohen, A. N., & Laws, J. M. 1992. An introduction to the ecology of the San Francisco Estuary. San Francisco Estuary Project.

Collins, J., J.G. Evens, & B. Grewell. 1994. A synoptic survey of the distribution and abundance of the California clapper rail, Rallus longirostris obsoletus, in the northern reaches of the San Francisco Estuary during the 1992 and 1993 breeding seasons. Technical Report to California Department of Fish and Game.

Cooper, N. J., Cooper, T., & Burd, F. 2001. 25 years of salt marsh erosion in Essex: Implications for coastal defence and nature conservation. Journal of Coastal Conservation, 7(1), 31-40.

DHI. 2011. Regional coastal hazard modeling study for North and Central San Francisco Bay. Final Draft Report. October 2011.

Dedrick. K.G. and L.T. Chu. 1993. Historical atlas of tidal creeks San Francisco Bay, California. Proceedings of the eighth symposium on coastal and ocean management (Coastal Zone 93). American Society of Engineers, New York, NY.

Doane, S.N. 1999. Shoreline changes in San Pablo Bay, California, Vanderbilt University Master's thesis, Nashville, Tennessee.

Ellis, M.Y. 1978. Coastal Mapping Handbook. Department of the Interior, U.S. Geological Survey and U.S. Department of Commerce, National Ocean Service and Office of Coastal Zone Management, U.S. GPO, Washington, D.C.

Fagherazzi, . 2013. The ephemeral life of a salt marsh. *Geology* 41, no. 8 (2013): 943-944.

Fagherazzi, S., Kirwan, M. L., Mudd, S. M., Guntenspergen, G. R., Temmerman, S., D'Alpaos, & Clough, J. 2012. Numerical models of salt marsh evolution: Ecological, geomorphic, and climatic factors. *Reviews of Geophysics, 50*(1).

Francalanci, S., Solari, L., Cappietti, L., Rinaldi, M., and Federici, G.V. 2011. Experimental observation on bank retreat of salt marshes. *Proc. River, Coastal and Estuarine Morphodynamics, RCEM 2011*, 543-551.

Gallinas Creek, October 1974 (photo #2006.34.18264.sm), Brady Collection, courtesy of Marin History Museum.

Goals Project. 1999. The baylands ecosystem habitat goals: A report of habitat recommendations Prepared by the San Francisco Bay Area Wetlands Ecosystem Goals Project, U.S. Environmental Protection Agency, San Francisco, California, and San Francisco Bay Regional Water Quality Control Board, Oakland, CA.

Goals Project. 2015. The baylands ecosystem habitat goals update for climate change: What we can do. The 2015 Science Update to the Baylands Ecosystem Habitat Goals prepared by the San Francisco Bay Area Wetlands Ecosystem Goals Project. California State Coastal Conservancy, Oakland, CA.

Gilbert, G.K. 1917. Hydraulic mining debris in the Sierra Nevada, USGS Professional Paper 105.

Grossinger RM, Stein ED, Cayce K, et al. 2011. Historical wetlands of the southern California coast: an atlas of U.S. Coast Survey t-sheets, 1851-1889. SFEI contribution #586, SCCWRP technical report #589. San Francisco Estuary Institute, Oakland, CA.

Gunnell, J. R., Rodriguez, A. B., & McKee, B. A. 2013. How a marsh is built from the bottom up. Geology, 41(8), 859-862.

Hapke, C. et al. 2006. National assessment of shoreline change part 3: Historical shoreline change and associated coastal land loss along sandy shorelines of the California coast. USGS Open File Report 2006-1219.s

Hapke, C. J., Himmelstoss, E. A., Kratzmann, M. G., List, J. H., & Thieler, E. R. 2011. National assessment of shoreline change; historical shoreline change along the New England and Mid-Atlantic coasts (No. 2010-1118). US Geological Survey.

Himmelstoss, E.A. 2009. DSAS 4.0 Installation Instructions and User Guide. In: Thieler, E.R., Himmelstoss, E.A., Zichichi, J.L., and Ergul, Ayhan. 2009 Digital Shoreline Analysis System (DSAS) version 4.0 — An ArcGIS extension for calculating shoreline change: U.S. Geological Survey Open-File Report 2008-1278. *updated for version 4.3.

International Hydrographic Organization. 1994. Hydrographic Dictionary. Special Publication Number 32, Fifth Edition.

Jaffe, B. E., Smith, R. E., & Foxgrover, A. C. 2007. Anthropogenic influence on sedimentation and intertidal mudflat change in San Pablo Bay, California: 1856–1983. *Estuarine, Coastal and Shelf Science*, 73(1), 175-187.

Kirwan, M. L., & Murray, A. B. 2007. A coupled geomorphic and ecological model of tidal marsh evolution. *Proceedings of the National Academy of Sciences*, 104(15), 6118-6122.

Kirwan, M. L., Guntenspergen, G. R., D'Alpaos, A., Morris, J. T., Mudd, S. M., & Temmerman, S. 2010. Limits on the adaptability of coastal marshes to rising sea level. Geophysical Research Letters, 37(23).

Kirwan, M. L., Murray, A. B., Donnelly, J. P., & Corbett, D. R. 2011. Rapid wetland expansion during European settlement and its implication for marsh survival under modern sediment delivery rates. Geology, 39(5), 507-510.

Lacy, J. R., & Hoover, D. J. 2011. Wave exposure of Corte Madera Marsh, Marin County, California—a field investigation (No. 2011-1183). USGS.

Lawson JS, Welker PA. 1887. Re-survey of San Pablo Bay, California, sheet no. 4. U.S. Coast and Geodetic Survey (USCGS). Courtesy of National Oceanic and Atmospheric Administration.

Lewicki, M., & McKee, L.J. 2010. New methods for estimating annual and long-term suspended sediment loads from small tributaries to San Francisco Bay. IAHS-AISH publication, 121-125.

Malamud-Roam, F., & Ingram, B. L. 2004. Late Holocene ⬜ 13 C and pollen records of paleosalinity from tidal marshes in the San Francisco Bay estuary, California. Quaternary Research, 62(2), 134-145.

Malamud-Roam, F., Dettinger, M., Ingram, B. L., Hughes, M. K., & Florsheim, J. L. 2007. Holocene climates and connections between the San Francisco Bay estuary and its watershed: a review. San Francisco Estuary and Watershed Science, 5(1).

Marani, M., d'Alpaos, A., Lanzoni, S., & Santalucia, M. 2011. Understanding and predicting wave erosion of marsh edges. *Geophysical Research Letters*,*38*(21).

Mariotti, G., & Fagherazzi, S. 2010. A numerical model for the coupled long⬜term evolution of salt marshes and tidal flats. *Journal of Geophysical Research: Earth Surface (2003–2012)*, *115*(F1).

McLoughlin, S. M., Wiberg, P. L., Safak, I., & McGlathery, K. J. 2014. Rates and forcing of marsh edge erosion in a shallow coastal bay. *Estuaries and Coasts*, 1-19.

McKee, L. J., Lewicki, M., Schoellhamer, D. H., & Ganju, N. K. 2013. Comparison of sediment supply to San Francisco Bay from watersheds draining the Bay Area and the Central Valley of California. Marine Geology, 345, 47-62.

Miller, A. 1967. Smog and Weather—The Effect of the San Francisco Bay on the Bay Area Climate. San Francisco Bay Conservation and Development Commission.

Möller, I., & Spencer, T. 2002. Wave dissipation over macro-tidal saltmarshes: Effects of marsh edge typology and vegetation change. *Journal of Coastal Research*, *36*(1), 506-521.

Mudd, S. M. 2011. The life and death of salt marshes in response to anthropogenic disturbance of sediment supply. Geology, 39(5), 511-512.

NOAA Special Publication NOS CO-OPS 1. 2000. Tidal Datums and Their Applications. U.S. Department of Commerce. Silver Spring, MD. Accessed April 4, 2008, at http://tidesandcurrents.noaa.gov/publications/tidal_datums_and_their_applications.pdf.

Odum W.E. 1990. Internal processes influencing the maintenance of ecotones: do they exist? In The Ecology and Management of Aquatic-Terrestrial Ecotones. Edited by Naiman and Décamps. The Parthenon Publishing Group. New Jersey, N.J.

Patrick, W. H. & DeLaune R.D. 1990. Subsidence, accretion, and sea level rise in south San Francisco Bay marshes. Limnology and Oceanography, 35(6), 1389-1395.

Pedersen, J. B., & Bartholdy, J. 2007. Exposed salt marsh morphodynamics: an example from the Danish Wadden Sea. *Geomorphology*, *90*(1), 115-125.

Pethick, J. S. 1992. Saltmarsh geomorphology. Saltmarshes: morphodynamics, conservation and engineering significance, 41-62.

Prahalad, V., Sharples, C., Kirkpatrick, J., & Mount, R. 2015. Is wind-wave fetch exposure related to soft shoreline change in swell-sheltered situations with low terrestrial sediment input? Journal of Coastal Conservation, 19(1), 23-33.

Rodgers AF. 1856. San Francisco Bay, California, sheet no. 22. U.S. Coast Survey (USCS). Courtesy of National Oceanic and Atmospheric Administration.

Schoellhamer, D. H. 2011. Sudden clearing of estuarine waters upon crossing the threshold from transport to supply regulation of sediment transport as an erodible sediment pool is depleted: San Francisco Bay, 1999. Estuaries and Coasts, 34(5), 885-899.

Schoellhamer, D. H., Wright, S. A., & Drexler, J. Z. 2013. Adjustment of the San Francisco estuary and watershed to decreasing sediment supply in the 20th century. Marine Geology, 345, 63-71.

Schwimmer, R. A., & Pizzuto, J. E. 2000. A model for the evolution of marsh shorelines. *Journal of Sedimentary Research*, 70(5).

Schwimmer, R. A. 2001. Rates and processes of marsh shoreline erosion in Rehoboth Bay, Delaware, USA. Journal of Coastal Research, 672-683.

Shalowitz AL. 1964. *Shore and sea boundaries, with special reference to the interpretation and use of Coast and Geodetic Survey data, United States.* U.S. Department of Commerce, Coast and Geodetic Survey. [Washington]: Government Printing Office.

Stralberg, D., Brennan, M., Callaway, J. C., Wood, J. K., Schile, L. M., Jongsomjit, D., M Kelly, M., Parker, V.T., Crooks, S. 2011. Evaluating tidal marsh sustainability in the face of sea-level rise: a hybrid modeling approach applied to San Francisco Bay. *PloS one*, 6(11), e27388.

Swanson, K. M., Drexler, J. Z., Schoellhamer, D. H., Thorne, K. M., Casazza, M. L., Overton, C. T., Takekawa, J. Y. 2014. Wetland accretion rate model of ecosystem resilience (WARMER) and its application to habitat sustainability for endangered species in the San Francisco Estuary. *Estuaries and coasts*, 37(2), 476-492.

Temmerman, S., Bouma, T. J., Govers, G., Wang, Z. B., De Vries, M. B., & Herman, P. M. J. 2005. Impact of vegetation on flow routing and sedimentation patterns: Three-dimensional modeling for a tidal marsh. Journal of Geophysical Research: Earth Surface (2003–2012), 110(F4).

Thieler, E.R., Himmelstoss, E.A., Zichichi, J.L., and Ergul, Ayhan. 2009. Digital Shoreline Analysis System (DSAS) version 4.0— An ArcGIS extension for calculating shoreline change: U.S. Geological Survey Open-File Report 2008-1278.

U.S. Coast and Geodetic Survey, Raymond Stanton Patton. 1917. United States coast pilot: Pacific coast: California, Oregon, and Washington Edition 3. G.O.P., University of Minnesota.

van der Wal, D., & Pye, K. 2004. Patterns, rates and possible causes of saltmarsh erosion in the Greater Thames area (UK). *Geomorphology*, 61(3), 373-391.

Van der Wegen, M., & Jaffe, B. E. 2013. Towards a probabilistic assessment of process-based, morphodynamic models. *Coastal Engineering*, 75, 52-63.

Van Eerdt, M. M. 1985. Salt marsh cliff stability in the Oosterschelde. *Earth Surface Processes and Landforms*, 10(2), 95-106.

Veloz, S., N. Elliott, D. Jongsomjit. 2013. Adapting to sea level rise along the north bay shoreline. A report to the North Bay Watershed Association. Point Blue Conservation Science.

Walters, R. A., & Gartner, J. W. 1985. Subtidal sea level and current variations in the northern reach of San Francisco Bay. *Estuarine, Coastal and Shelf Science*, 21(1), 17-32.

Wright, S. A., & Schoellhamer, D. H. 2004. Trends in the sediment yield of the Sacramento River, California, 1957–2001. San Francisco Estuary and Watershed Science, 2(2).

Zoulas, J., unpublished, Summary of Corte Madera Shoreline Study, Data report prepared for the San Francisco Bay Conservation and Development Commission.